AF657813

# Beyond Newton and Archimedes

# Beyond Newton and Archimedes

Ajay Sharma

Viva Books
New Delhi | Mumbai | Chennai | Kolkata
Bengaluru | Hyderabad | Kochi | Guwahati

Copyright © Ajay Sharma

**First Indian Edition 2014**
**Reprinted 2015**

**VIVA BOOKS PRIVATE LIMITED**

- 4737/23 Ansari Road, Daryaganj, New Delhi 110 002
  E-mail: vivadelhi@vivagroupindia.net, Tel. 42242200
- 76, Service Industries, Shirvane, Sector 1, Nerul, Navi Mumbai 400 706
  E-mail: vivamumbai@vivagroupindia.net, Tel. 27721273, 27721274
- Megh Towers, Old No. 307, New No. 165, Poonamallee High Road, Maduravoyal
  Chennai - 600 095 E-mail: vivachennai@vivagroupindia.net, Tel. 23780991, 23780992
- B-103, Jindal Towers, 21/1A/3 Darga Road, Kolkata 700 017
  E-mail: vivakolkata@vivagroupindia.net, Tel. 22836381, 22816713
- 7, Sovereign Park Apartments, 56-58, K. R. Road, Basavanagudi, Bengaluru 560 004
  E-mail: vivabangalore@vivagroupindia.net, Tel. 26607409, 26607410
- 101-102 Mughal Marc Apartments, 3-4-637 to 641, Narayanguda, Hyderabad 500 029
  E-mail: vivahyderabad@vivagroupindia.net, Tel. 27564481, 27564482
- Beevi Towers, First Floor, SRM Road, Kaloor, Kochi 682 018, Kerala
  E-mail: vivakochi@vivagroupindia.net, Tel: 0484-2403055, 2403056
- 232, GNB Road, Beside UCO Bank, Silpukhuri, Guwahati 781 003
  E-mail: vivaguwahati@vivagroupindia.net, Tel: 0361-2666386

*www.vivagroupindia.com*

*Published by arrangement with*
Cambridge International Science Publishing
7 Meadow Walk, Great Abington,
Cambridge CB21 6AZ, UK

All rights reserved. No part of this book may be reproduced or transmitted in any form or by any means now known or to be invented, electronic or mechanical, including photocopying, recording or by any information storage or retrieval system without written permission from the publisher.

*This edition is for sale in the Indian subcontinent only. Not for export elsewhere.*

ISBN : 978-81-309-2674-2

Published by Vinod Vasishtha for Viva Books Private Limited, 4737/23, Ansari Road, Daryaganj, New Delhi - 110 002. Printed and bound by Thomson Press, New Delhi.

## Contents

# Preface

*The spirit or essence of science is lighting one lamp from the other.*

Like God, Isaac Newton's wisdom is indescribable, yet he was too a human being. Newton is the greatest scientist of all time due to his matchless imagination and unsurpassed intuition, which of course proved consistent with experimental findings in due course of time. It is irony that the second law of motion ($F = ma$), which we read and teach as Newton's Second Law of motion, was not given by him. But it is a lapse of the others. It can be checked from his immortal masterpiece, *the Principia*, translated by Andrew Mott and published in 1729, two years after legend's death. The three laws of motion are described on page 19. Newton presented the law in an entirely different way. Further, the basis for Newton's first and second law of motion was provided by Aristotle (384–322 BC), Johannas Philiponus (470-570BC), Jean Buridan (1300–1358) and Galileo Galilei (1564–1642). The third law of motion is solely Newton's creation. An attempt had been made to understand the most fundamental laws of science.

Archimedes is one of the greatest mathematicians in the history of science. The principle known after his name has been taught all over the world for 2263 years. There is no other example of this kind in science. However, the equations based upon Archimedes do not account for the shape of the body, the viscosity of the medium, etc; it is the limitation of the principle. It can be explained if Archimedes principle is generalized, i.e. *the upthrust is regarded as proportional to the weight of the fluid displaced.* According to the Archimedes principle, the *upthrust is equal to the weight of the fluid displaced.* Also, under some particular conditions, the Archimedes principle predicts that the volume of the medium filling a balloon becomes indeterminate or meaningless, i.e. $V = 0/0$. The generalized form of the Archimedes principle explains the same as the exact volume $V$ is obtained. It can be confirmed with sensitive experiments. For the first time in the history of science a generalized theory of rising, falling and floating bodies is developed. The theories of Aristotle have been mysteriously sometime accepted and sometime rejected. But here the facts are presented which yield acceptance to Aristotle's assertion about falling bodies especially in the region of validity of Stokes Law. The mercury barometer was constructed by Italian scientist Evangelista Torricelli in 1644. In the book, it is also stressed that a water barometer, a glycerin barometer, etc. should be constructed. For the water barometer, the length of tube must be more than 10.3 m, so it would be a scientific, spectacular and historical experiment. It would definitely highlight many facts about fluid dynamics.

The main ideas, which are presented in the book, struck me in 1982, at the age of 19, as a student of BSc.II. At that young age ideas were carried away by self intuitions only. That was advantage or disadvantage, only God knows. During MSc. (Physics) studies, Professor Abdus Salam Pakistani Nobel Laureate, founder and Director of the International Center for Theoretical Physics encouraged very much. But the main advantage is that I got considerable time for such innovative activities. Research is time consuming but research of established topics takes more time and is risky. In the process many research papers were published in journals and presented in various conferences including in the USA and England.

Some people have years and years for criticism but really no time for constructive suggestions. The knowledge is infinite and suggestions from one and all are humbly solicited, so that the subject matter which is just introduced may be improved. Thus, science is like lighting one lamp from the other.

I am indebted to my critics and promoters within the house, e.g. my late mother Mrs. Kala Rani, the poet father late Mr. Dev Dutt Sharma and the wife Anjana. Also the daughter Abhilasha read the manuscript and Abhinav helped in computer work Otherwise, it would have been difficult to maintain enthusiasm over three decades on such topics. Also diligent efforts of Professor Shashi Kant Lomesh, Chairman of the Department of Chemistry at the HP University Summer Hill Shimla, are gratefully acknowledged. Finally, I am highly thankful to publisher Victor Riecansky for his untiring efforts in bringing out the book.

Suggestions and criticism are solicited at ajay.pqrs@gmail.com.

*Ajay Sharma, Assistant Director of Education, Shimla*

1

# 2360 Years Old Aristotle's Assertion Revalidated by Stokes Law

## First Glimpse

- ✓ According to Aristotle's assertion of falling bodies, 'heavier body falls down quickly'.
  It got immediate support from the fact that a stone falls more quickly than a straw. Mathematically, **falling tendency $\propto$ mass of the body.**
- ✓ In 1851, Stokes put forth that under certain conditions the bodies attain constant velocity. According to relevant mathematical equations,

  **Average velocity $\propto$ mass of body**

- ✓ It was contradicted by Galileo [1564–1642] after 2000 years of use stating that all bodies fall with the same acceleration i.e. travel equal distances in equal intervals of time. It got immediate support from the fact that a ten pound shot and one pound shot fall at same time in air. Thus, 'all bodies fall equal distances from equal intervals of time.'
  Mathematically

  $$S = \frac{1}{2}at^2$$

  where $a$ is acceleration.
- ✓ Comparing Aristotle's assertion and Stokes law, it is obvious that both have similar forms. Thus, Aristotle's assertion is as useful as Stokes' law in the case of the falling bodies.

✓ If the resultant weight is regarded as falling tendency, then Falling tendency $\propto$ mass
Thus, in this case also Aristotle's assertion is justified. Acceleration is required to calculate the distance travelled by the body.

## 1.0. Introductory Mathematical Background of Aristotle's Assertion

When people began to think about motion of bodies on the Earth, the different opinions emerged. Ideas of various thinkers appeared in antiquity even before Christ. There is no doubt in the fact that most of the ideas may have been destroyed or not recorded or found scientifically unsound. However, the ideas of Aristotle stood to the test of arguments and time and remained prevalent for about 200 years. Thus, Aristotle may be regarded as the first physicist. The simplicity is hallmark of Aristotle's perceptions.

Millions might have seen the slow motion of straw in the atmosphere. But it was Aristotle [1] who compared the motion of a falling stone in air with the straw (a bit of paper) to conclude that the stone (heavier body) falls more quickly than the straw (lighter body).

Aristotle had explained how and why objects fall. He said that every object has a natural place in the universe. If it is moved from the natural place and released, it will move back. A heavy body is one that would fall on the Earth, if released in air. Alight body would rise. These explanations stood for 2000 years until the great scientist finally contradicted them by means of experiments and rational arguments in air (not in dense fluids). Currently, all arguments are tested in view of experiments in dense liquids and new ideas emerge.

So Aristotle's assertion has been contradicted about after 2000 years of continuous use. To explain the critical observation of different motion of a sheet of paper and a metallic ball (which supports Aristotle's assertion), Galileo [1564–1642] held air resistance responsible. But Galileo did not calculate the air resistance due to lack of mathematical background existing at that time. Also, he did not suggest any quantitative equation. But it became possible to calculate the upthrust in 1685 when Newton published *The Principia* and defined acceleration due to gravity $g$ and weight [2].

## 1.1. Aristotle (384–322 BC) and his assertion

Hipparchus (190–120 BC), Claudius Ptolemy (85–165 AD) and Aristotle (384 BC–322 BC) were three great Greeks in the ancient time. Hipparchus was born in Turkey and is considered to be the greatest observational astronomer of ancient Greece. The last great astronomer of the ancient world was Claudius Ptolemaeus (known as Ptolemy to posterity). Aristotle is the third member of this trio born in 384 BC in Stagira, in Thrace, at the northern end of the Aegean, near Macedonia. Aristotle's father was the family physician of King Philip of Macedonia. He was trained first in medicine, and then in 367 he was sent to Athens to study philosophy with Plato. He stayed at Plato's Academy until about 347. Though a brilliant pupil, Aristotle opposed some of Plato's teachings, and when Plato died, Aristotle was not appointed the head of the Academy.

He returned to Macedonia in 338 BC to tutor Alexander the Great. After Alexander conquered Athens, Aristotle returned to Athens and set up a school of his own, known as the Lyceum. After Alexander's death, Athens rebelled against the Macedonian rule, and Aristotle's political situation became precarious. To avoid being put to death, he fled to the island of Euboea, where he died soon after. Aristotle is said to have written 150 philosophical treatises. The 30 that survive touch on an enormous range of philosophical problems, from biology and physics to morals to aesthetics to politics. But here we will be only concerned about the motion of bodies.

Regarding the motion of bodies, his laws are important at least on two reasons. Firstly, they influenced scientific thought for nearly two thousand years, i.e. were taught with great faith by the philosophers and teachers. Secondly, they provide a common sense framework that most people would agree that perceptions or laws make sense.

To make his point of view clear, Aristotle categorised motions in two ways, i.e. natural motion and violent motion. Here by natural motion it is meant that motion is not under the influence of external forces i.e. occurs naturally. The examples of natural motion are:

(*i*) A book lying on the table naturally remain at rest

(*ii*) If the book is dropped then it naturally falls down.

(*iii*) The puffs of smoke naturally rise.

## 1.2. Violent motion or forced motion

Aristotle termed the violent motion as motion under the influence of external force(s).

'One of the fundamental prepositions of Aristotelian philosophy is that there is no effect without cause.'

Applied to moving bodies, this preposition indicated that there is no motion without a force. For higher the speed of the body, the higher is the force to be applied. For the natural motion of heavy objects falling to the Earth, Aristotle asserted that the

'*speed of fall was proportional to the weight*, and *inversely proportional to the density of the medium in which body falls.'*

Thus, speed is proportional to force and inversely proportional to resistance

Force $\propto$ speed

$$\text{Force} = \text{resistance} \times \text{speed} \tag{1.1}$$

Thus, Aristotle developed mathematical equations for his perceptions. It must be noted that Aristotle wrote about 150 treatises, and only 30 survive at the moment whereas in case of Newton's and Galileo's the intellectual contents are pretty safe comparatively. Here Aristotle's physics agrees with most people's common sense but modern scientists discard it.

Here the aim of discussion is to discuss the current status of Aristotle's assertion of falling bodies. It was contradicted by Italian scientific legend Galileo Galilei (1564–1642) in simple demonstration. The critical analysis of works of George Gabriel Stokes (1819–1903), indicated that Aristotle's assertion on falling bodies is true as long as the Stokes law is valid. Thus, Galileo's work is not useful in the fluids. Also, Galileo drew conclusions after conducting experiments in air and not in other fluids (like water) and bodies of different shapes were not taken in observations. It may be regarded as a limitation of the interpretation.

## *1.3. Aristotle's assertion about falling bodies*

According to Aristotle (384–322 BC), the heavier bodies have a higher tendency to fall than lighter one's [1]. Aristotle maintained that the Earth is the natural home of the material bodies, so the material bodies fall down to have the natural position. The assertion got the immediate support from the fact that a bit of paper falls more slowly than a metallic ball. Similarly, a metallic piece of a typical shape falls more quickly in water than a stone.

Aristotle stated the conclusion without mathematical equations, but now conclusions are drawn on the basis of exact mathematical equation. Thus, Aristotle's perception, i.e.

*'Heavier body falls quickly than lighter body'*

has to be changed to a mathematical equation. In Aristotle's days, the 'falling tendency of the body' can be understood to have been expressed in terms of 'average velocity'. The reason is that as in this regard the concept of acceleration was introduced by Galileo (1564–1642) and strengthened by Newton (1642–1727). So the above statement becomes

Falling tendency $\propto$ mass of falling body
Average velocity $\propto$ mass of falling body

$$v_{av} \propto m$$

or
$$v_{av} = Zm \tag{1.2}$$

where $Z$ is the coefficient of proportionality like Hubble's constant, or a coefficient, like the coefficient of thermal conductivity or viscosity. The value of $Z$ depends upon experimental conditions. Now, Eq. (1.2) relates the average velocity with the mass of the body, stating that higher the mass of the body more quickly it falls. If the falling body is lighter then its average velocity is lesser and it falls more slowly. Further, Aristotle simply meant the motion of bodies in air and did not mention anything about the medium in which the body falls. However, nowadays the falling tendency is expressed in terms of acceleration. Hence to draw conclusions quantitatively, the falling tendency has to be expressed in terms of acceleration.

There are kinematical equations in the existing physics but without clear evidences about the name or group of scientists who discovered them.

### *1.4. In air*

Consider a person is holding in hand two objects, e.g. a metallic ball (heavier body) and a piece of paper (lighter body). Aristotle, in fact, did not quantitatively distinguish between the lighter body and the heavier body. Also, Aristotle did not mention the medium in which the body is dropped. A person drops two objects simultaneously, say from a height of 30 feet. Now it is obvious that the heavier body (metallic ball) falls quickly than the lighter body (piece of paper).

Thus, Aristotle's assertion that the heavier body (metallic ball) falls more quickly than the lighter body (straw) is experimentally confirmed in air. It can never be denied, even now. Aristotle meant his assertion for such situations. There is no mention for which medium Aristotle meant his thesis. So it is discussed in air and water, i.e. in fluids for completeness. Scientific inferences are not drawn only looking at one side of picture.

### *1.5. In water*

Further for completeness consider an observation in water. Let there be a steel sphere of mass 5 kg and another flat body of mass 50 gm. Both bodies are dropped in a tank of water of sufficiently large dimensions. The heavier body (sphere of mass 5 kg) reaches the bottom more quickly than the small body (flat body of mass 50 gm). So the heavier body falls more quickly than the lighter body. Thus, in this case it is also found that Aristotle's assertion is experimentally justified.

Thus, Aristotle's assertion that the heavier body falls more quickly than the lighter body is experimentally confirmed in water also. The same can also be justified in other denser liquids. Thus, so far Aristotle's assertion

'*Heavier body falls more quickly than the lighter body, it appears to be true in water in some cases.*'

In the above cases the difference between the masses of the two bodies is quite large.

### 2.0. Galileo's contradiction of Aristotle's assertion

Galileo was born in Pisa, Italy, on February 15, 1564 and died in 1642. His father, Vincenzo Galilei, was a musician. His family belonged to the nobility but was not rich. In 1581, Galileo began studying at the University of Pisa, where his father hoped he would study medicine. In 1592, Galileo was appointed a professor of mathematics at the University of Padua. Galileo made his first telescope in 1609 and studied the motion of planets. Galileo did not marry but he had three daughters from a brief relationship with Marina Gamba, a woman he met on one of his many trips to Venice.

His discoveries proved the Copernican system which states that the Earth and other planets revolve around the Sun. Galileo's belief in the Copernican system eventually got him into trouble with the Catholic Church. The Inquisition was a permanent institution in the Catholic Church charged with the eradication of heresies. Galileo faced the wrath of inquisition for his bold, frank and true doctrines. He was warned not go ahead with his views on the Copernican system in public in 1616.

In 1632, Galileo published a book, *Dialogue Concerning the Two Chief World Systems*. In the book, Galileo had preached that the Sun is the centre of the Universe, and the Earth revolves around the Sun, which conflicted with the Holy Bible. Consequently, he was called to Rome in 1633 to face the Inquisition again. Galileo was found guilty of heresy for his book The Dialogue, and was sent to his home near Florence where he was to be under house arrest for the remainder of his life. In 1638, the Inquisition allowed Galileo to move to his home in Florence. By that time he was totally blind. In 1642, Galileo died at his home outside Florence. His daughter cared for him in the last days of his life.

Galileo, the boldest scientist who ever walked on the Earth, who proved that

'The Earth revolves around the Sun'

also critically analyzed the Aristotle's assertion about the falling body. Galileo's two books are famous and shook the very foundations of science

(*i*) The *Dialogue Concerning the Two Chief World Systems*: this book was published in 1632 and established that the Earth revolves around the Sun. The ideas contradicted the Holy Bible and Galileo faced fury of the Inquisition and was imprisoned. The book was immediately banned.

(ii) The *Discourses and Mathematical Demonstrations Relating to Two New Sciences*: this book was published in 1638 about four years before the death of Galileo. In this book Galileo contradicted the Aristotle's assertion about falling bodies.

Contrary to Aristotle's assertion, Galileo maintained that all bodies fall downward at the same rate, i.e. travel equal distances in equal intervals of time. It is precisely true in vacuum. The experiments involving bodies of different shapes in denser media may lead to different results.

### *2.1. Mathematical basis of Galileo's conclusions*

Mathematically all bodies fall with the same acceleration. Acceleration is understood as

*the rate of change of velocity.*

Acceleration = [Terminal velocity – Initial Velocity] / [Final time – Initial time] $[T_f - T_i]$

$$a = [V_f - V_i] / [T_f - T_i] \tag{1.3}$$

There are kinematical equations in the existing physics but without clear evidences about the name or group of scientists who discovered them.

The distance travelled by a body in certain time is related with uniform velocity and acceleration. We have different equations to determine the distance travelled when the body moves with uniform velocity and acceleration.

### *2.2. Distance travelled when the body moves with uniform velocity*

Let a body moves with uniform velocity $v$ and cover distance $S$ in time $t$. Then, the distance travelled is given by

$$S = vt \tag{1.4}$$

Equation (1.4) is applicable when the body moves with uniform velocity. Also, when the body moves with uniform velocity, then uniform velocity and average velocity are both the same. In this case, the acceleration of the body is zero.

### *2.3. Distance travelled when the body moves with uniform acceleration*

The motion of the body is said to be accelerated, if the velocity increases regularly at regular intervals of time. There are few equations which relate the distance travelled with acceleration. But here the simplest equation is used to calculate the distance travelled. It can be calculated with the help of the kinematic equation

$$S = ut + \frac{1}{2}at^2 \tag{1.5}$$

where $S$ is the distance travelled, $u$ is initial velocity, $a$ is acceleration, and $t$ is time. If the body starts from the rest, i.e. $u = 0$, then Eq. (1.5) becomes

$$S = \frac{1}{2}at^2 \tag{1.6}$$

According to Eq. (1.6), the distance travelled by the body depends upon its acceleration. If bodies of different masses (heavier and lighter) fall with the same acceleration, then they travel equal distances in equal time intervals. It is not consistent with Aristotle's assertion of falling bodies that the heavier body falls more quickly than the lighter one. Moreover, Aristotle (384 BC–322 BC) did not give any mathematical equation to justify his assertion.

## *2.4. Galileo's experimental demonstration*

It is alleged that Galileo performed a famous experiment from the Leaning Tower of Pisa. This is considered an apocryphal tale, its only source being Galileo's secretary. The Leaning Tower of Pisa has a height of 55.86 m and has 294 steps. However, there is no proof in favour of this fact that the experiment has been ever done from the Leaning Tower of Pisa. Also, there is no data to support this fact. Such experiment can be performed from any height by any one.

It is believed that he climbed up the Tower with one pound and 10 pound shots. Then he threw both shots simultaneously from the same height in front of a gathering. Obviously, both the one pound and 10 pound shots struck the ground simultaneously. (However, a stone and a straw fall on the ground at different times from the same Leaning Tower of Pisa). Mathematically,

$$S = \frac{1}{2}at^2 \tag{1.6}$$

where $S$ is the distance travelled, $a$ is acceleration and $t$ is time. Thus, if the acceleration is the same then the bodies will fall through equal distances in equal intervals of time. Hence mathematically it is being said that all bodies fall with the same acceleration. Now it is confirmed that all bodies fall down with the same acceleration equal to 9.8 m/s$^2$. Thus,

$$S = \frac{1}{2}gt^2 = \frac{1}{2}9.8t^2 = 4.9 \text{ m} \tag{1.7}$$

The distance travelled by the heavy body of 10 pounds in 1 s = 4.9 m (1.8)

The distance travelled by the light body of 1 pound in 1 s = 4.9 m (1.9)

Thus, in practical language all bodies travel equal distances in equal intervals of time. However, Eqs. (1.8–1.9) do not hold good for the stone and the straw (tremendous difference between masses, this aspect can be quantitatively studied for final conclusions).

According to Aristotle's assertion on falling bodies, the heavier body (10 pound shot) should fall more quickly than the lighter body (one pound shot). Also, in Aristotle's days (384–322 BC) there were no mathematical equations to check any deduction or preposition. Thus, Aristotle's assertion is contradicted here qualitatively. But it is added here that Galileo should have also conducted experiments involving bodies of different shapes in water or other fluids. Thus, a conclusion should be drawn considering the wide range of experimental parameters. It is not scientific to draw conclusions from the experiments conducted in the limited range and generalize the results for all cases.

## 2.5. *In vacuum*

In 1987, Niebauer, Mchugh and Faller dropped various bodies of different masses in vacuum and measured the speed with the most sophisticated equipment. In other words, all the bodies fell in vacuum with the same rate or acceleration. All bodies travelled equal distances in equal intervals of time. This conclusion is drawn on the basis of Eq. (1.7). Thus, under this condition Galileo's doctrine can be perfectly obeyed.

So Aristotle's assertion has been contradicted after about 2000 years of continuous use. To explain the critical observation of different motion of a sheet of paper and a metallic ball (which supports Aristotle's assertion), Galileo [1564–1642] held air resistance responsible. The different results are due to different resistances of media (water and air) which the falling bodies experience. In Galileo's time, there was no mathematical equation to calculate it.

But Galileo did not calculate the air resistance due to the lack of mathematical background existing at that time. But it became possible to calculate the upthrust in 1685 when Newton published *The Principia* and defined acceleration due to gravity $g$ and weight [2]. However some qualitative or empirical methods could have been devised by Galileo, depending upon mass, shape, size etc. of the body and the density of the medium.

In vacuum, the 1 lb shot and the 10 lb shot are found to fall with the same acceleration, so

$$S = \frac{1}{2}gt^2 \quad = \frac{1}{2}9.8t^2 \ = 4.9 \text{ m} \tag{1.7}$$

In vacuum a stone and a straw are found to fall with the same acceleration then

$$S = \frac{1}{2}gt^2 \quad = \frac{1}{2}9.8t^2 \ = 4.9 \text{ m} \tag{1.7}$$

So Galileo's demonstration is precisely true for vacuum.

## 3.0. Merits and demerits of Aristotle's assertion and Galileo's demonstration

Aristotle and Galileo did not specifically mention any medium in which the body is observed. Normally, it is considered to be air.

(*i*) So under some conditions in air and liquids Aristotle's assertion appear to be true. It can never be denied. It is supported by the motion of different masses and shapes. If some specific experiments are conducted then concrete conclusions are drawn. A body of mass 50 kg (spherical) falls more quickly in water than a body of 5 gm mass (flat or distorted), like the stone falls more quickly in air than in water. However, Aristotle's assertion cannot correctly explain the motion of bodies in vacuum.

(*ii*) The great Galileo conducted experiments in air, but not in liquids which would have completed the observations. In liquids, Galileo would have encountered the rising bodies as well, so this may have led to a new theory on rising and falling bodies. Thus, no theory should be developed looking at either half or partial phenomena. Galileo definitely commented that it is the air resistance which affects the results. But he did not empirically or qualitatively gave any method to calculate the air resistance. In vacuum all bodies fall equal distances in equal times, as in vacuum the resistive forces are zero.

(*iii*) In vacuum Galileo's doctrine is completely true that all bodies travel equal distances in equal intervals of time. In vacuum there is no air resistance. When the body falls in the partial vacuum or air or liquids then it experiences an upthrust, depending upon density. But the distance travelled is related with acceleration as in Eq. (1.5).

(*iv*) The Archimedes principle defines the upthrust (the weight of the fluid displaced by the body). But at Galileo's time upthrust could have not been calculated mathematically due to a lack of terminology. In Galileo's days, the acceleration due to gravity was not defined. With the help of acceleration due to gravity $g$ the weight and upthrust are both defined. The acceleration due to gravity was defined in 1685 when Newton published the law of gravitation in the *Principia*. If the existing phenomena are critically studied then phenomena regarding falling and rising bodies are not explained quantitatively, in spite of the fact that upthrust, weight and resultant weight are all defined.

**Table I.** Ranges of validity of Aristotle's assertion (heavier body falls quickly than lighter) and Galileo's doctrine (all bodies fall down equal distances in equal intervals of time)

| Sr. No | Perception | Air | Liquid | Vacuum |
|---|---|---|---|---|
| 1 | Aristotle's Assertion | Partially valid A stone falls quickly than straw | Partially valid A 50 kg (spherical) ball falls more quickly than 5 g (flat) | Invalid All bodies reach bottom at same time |
| 2 | Galileo's Doctrine | Completely valid if air resistance is zero in vacuum | Not obeyed in many cases | Valid completely |

## 4.0. Mathematical equations based upon Archimedes principle, and falling bodies

Galileo maintained that all bodies travel equal distances in equal intervals of time. But it is precisely true in vacuum only. In air it was argued that the stone falls quickly than the straw, for the reason that the straw experiences more air resistance hence falls more slowly. However, Galileo did not calculate the air resistance experienced by the stone and the straw. So this explanation is qualitative only. Mathematically, the distance travelled is related with acceleration, so the acceleration of the body has to be calculated. In vacuum, all bodies fall with constant acceleration which is equal to 9.8 $m/s^2$. However, the resultant acceleration is less in fluids. It can be calculated with the help of the Archimedes principle. But the mathematical equations were developed after 1935 years of enunciation of the principle when Newton defined acceleration due to gravity, *g* in the *Principia*. This is the importance of the Archimedes principle in such phenomena.

Archimedes stated about 2265 years ago that

'*when a body is wholly or partially immersed in a fluid at rest then it experiences upthrust and its weight decreases. The decrease in weight is equal to the weight of the fluid displaced.*'

It is believed (but without documentary proof) that King Heiron once asked the mathematician Archimedes to find out whether his crown was made of pure gold or not. The problem baffled Archimedes for a long time. The legend is that once Archimedes was taking his bath in a tub, found that his weight has decreased in water and he experienced an upthrust.

From this observation Archimedes is believed to have invented methods for checking the purity of King's crown. But there are no authentic proofs in the existing literature that this incidence had actually happened in the history of science. So it based upon just hearsay. It is similar to Galileo's demonstration from the Leaning Tower of Pisa, both are based upon hearsay. However, many people knew even in remote areas that when a body is immersed in water it loses its weight, even without knowing about Archimedes and his principle. Galileo concluded that the resistance of the medium is responsible for this.

### 4.1. Mathematical equations based on Archimedes principle

The mathematical equations based upon the Archimedes principle became feasible in 1685 after 1935 years of the enunciation of the principle.

Mathematically, the weight of a body is given by

$$\begin{aligned} W &= \text{Mass} \times \text{acceleration due to gravity} \\ &= Mg \end{aligned} \tag{1.10}$$

The equation for weight follows from the second law of motion which in the mathematical form can be expressed as

$$F = Ma \tag{1.11}$$

where $F$ is the external impressed force, $M$ is mass and $a$ is the acceleration produced in the body.

In free fall, the acceleration of a body is 9.8 m/s$^2$ which is denoted by $g$. Now, replacing acceleration $a$ by the acceleration due to gravity $g$, we get

$$F = Mg \tag{1.12}$$

This force is denoted by weight $W$ as

$$W = Mg \tag{1.10}$$

The mass, density and volume of the body are related with each other as,

$$\text{Mass of body} = \text{density of body} \times \text{volume of body} = D_b V \tag{1.13}$$

$$\text{Weight, } W = D_b V g \tag{1.10}$$

It acts downwards towards the surface of the Earth. When a body is immersed in a fluid then it experiences an upthrust (regarded as the resistance for the body to fall down) which is given by

$$\begin{aligned} U &= \text{weight of fluid displaced} \\ &= \text{mass of fluid displaced} \times \text{acceleration due to gravity} \\ &= \text{density of medium} \times \text{volume of body} \times \text{acceleration due to gravity} \\ &= D_m V g \end{aligned} \tag{1.11}$$

Now weight acts downwards and upthrust in the upward direction. The weight of the body in the fluid is known as the resultant weight.

Resultant weight = weight in vacuum – weight in medium or fluid

$$\text{Resultant weight} = D_b Vg - D_m Vg = (D_b - D_m)Vg$$

$$= (1-\frac{D_m}{D_b})VD_b g = (1-\frac{D_m}{D_b})Mg \qquad (1.12)$$

The resultant weight, i.e. the weight of the body in the fluid tends to accelerate the body downwards [3] but with a reduced magnitude as in fluids. The bodies have maximum (and constant) acceleration [4] equal to $g$ in vacuum.

In vacuum $D_m = 0$, then
Resultant weight = $(D_b V)g$ (1.13)

or acceleration due to gravity = $D_b Vg/ D_b V = g$ (1.14)

Hence all bodies fall with the same acceleration and travel equal distances in equal intervals of time. Thus, the equation is written for the medium with reduced to the acceleration due to gravity in vacuum.

### *4.2. Can resultant weight be a measure of tendency to fall?*

$$\text{Resultant weight} = (1-\frac{D_m}{D_b})VD_b g = (1-\frac{D_m}{D_b})Mg \qquad (1.15)$$

The weight is the force with which the body is attracted by the Earth towards its centre. Hence it can be expressed as the falling tendency. If bodies of the same density are studied in the same medium then $(1-\frac{D_m}{D_b})$ becomes constant, thus

The resultant weight or falling tendency $\alpha$ $M$

Thus, the heavier the body the more quickly it falls down (if the resultant weight is regarded as the falling tendency). This explanation suits Aristotle's assertion that the heavier body falls more quickly than the lighter body. In the existing literature this explanation is not given, hence it is neither accepted nor denied, just explained. However, the distance travelled is related with acceleration, thus acceleration is expressed as falling tendency. The acceleration can be calculated from the equation of resultant weight, i.e. Eq. (1.15). Similarly, acceleration can be regarded as the falling tendency of bodies, as it is associated with time and distance.

## 5.0. Re-validation of Aristotle's assertion by Stokes law

Aristotle justified his assertion in air. It is invalid in vacuum, however, it can be extended to denser fluids. If the assertion is extended to denser fluids then its utility becomes more clear. Galileo's demonstration is precisely valid in vacuum only, when the acceleration due to gravity $g$ is the same for all bodies.

### *5.1. Stokes law*

The viscosity is the property of the fluid (liquid or gas) by virtue of which internal friction comes into play when the fluid is in motion and opposes the relative motion between the fluid.

In everyday terms (and for fluids only), viscosity is 'thickness' or 'internal friction'. Thus, water is 'thin', having a lower viscosity, while honey is 'thick', having a higher viscosity. It is also called fluid friction. When a body falls in the viscous medium it experiences viscous force. The CGS unit of viscosity is dyne cm/s$^2$. The viscosity of water at 20°C is 0.01 Poise. The viscosity of blood at body temperature is 0.03 Poise.

Consider the motion of a body in a fluid. The result of this is that the backward dragging force (i.e. viscous force) comes into play, which opposes the motion of the body. This backward dragging force increases with increase with the velocity of the moving body. George Gabriel Stokes concluded that the backward dragging force $F$, acting on a small spherical body of radius $r$ moving through a medium of the coefficient of viscosity $\eta$ with velocity $v$, is known as the viscous force. The viscous force $F$ is given by

$$F = k\eta rv = 6\pi\eta rv. \tag{1.16}$$

Equation (1.16) derived by method of dimensions and the proportionality constant is determined experimentally. Stokes law is used to measure the electronic charge in Millikan's method, the speed of a falling drop, the speed of a person coming down with help of a parachute, etc. Here the same will be used in justifying that Aristotle's assertion in falling bodies is still useful up to some extent.

### *5.2. Constant velocity of small spheres*

Stokes put forth that small spherical bodies in a fluid fall with constant velocity under certain conditions. These conditions or postulates are

(*i*) when the body is a small sphere
(*ii*) falls through a medium of infinite extent (no disturbance is transmitted by the medium to the body).
(*iii*) medium is a viscous medium

Then the body falls with a constant velocity and its acceleration is zero. According to Galileo, the bodies fall with constant acceleration.

The above law is also experimentally justified. In 1910, Dr. Arnold experimentally confirmed the viscous force for a small sphere of Rose metal of radii 0.002 cm in water. It is reported by R.A. Millikan [5] in his book *The Electron*. Thus, the viscous force has experimental validity. This constant velocity of the small sphere falling under certain conditions can be calculated as below.

### *5.3. Calculation of terminal velocity or constant depending upon Stokes law*

When a small spherical body falls through a viscous medium at rest, under certain conditions, then three forces act on the sphere

(*i*) The weight of the body acting vertically downwards which is given by Eq. (1.10) as

$$W = D_b V g \tag{1.10}$$

(*ii*) The upthrust experienced by the immersed body in the medium of density $D_m$

$$U = D_m V g \tag{1.11}$$

It acts in the upward direction.

(*iii*) The viscous drag force acts in the direction opposite to which the body moves. Stoke's law,

$$F = 6\,\pi\,\eta\,r\,v. \tag{1.16}$$

is the opposing viscous drag which keeps on increasing with increasing velocity of the body. If the body falls downwards then the viscous force acts upwards.

As the body falls down, its downward speed keeps on increasing due to gravity. Consequently, the opposite viscous drag which acts in upward direction also increases. A stage is reached when the true weight of the body is just equal to the sum of the upward thrust and the upward viscous force. At this stage there is no net force to accelerate the body. Hence the body starts moving with constant velocity (zero acceleration). When the body attains constant velocity then

$$U + F = W$$

$$D_m Vg + 6\pi\eta rv = D_b Vg \tag{1.17}$$

In Stokes law, the small spherical body is considered thus its volume $V$ is

$$V = \frac{4}{3}\pi r^3 \tag{1.18}$$

Now substituting the value of the volume from Eq. (1.18) into Eq. (1.17) we get

$$D_m \frac{4}{3}\pi r^3 g + 6\pi\eta rv = D_b \frac{4}{3}\pi r^3 g$$

$$6\pi\eta rv = D_b \frac{4}{3}\pi r^3 g - D_m \frac{4}{3}\pi r^3 g$$

or $$6\pi\eta rv = (D_b - D_m) \frac{4}{3}\pi r^3 g \tag{1.17}$$

$$v = \frac{2r^2\left(D_b - D_m\right)g}{9\eta} \tag{1.18}$$

Further, Eq. (1.17) can be written as

$$v = \frac{\left(1 - \dfrac{D_m}{D_b}\right)Mg}{6\pi\eta r} \tag{1.19}$$

The Eqs. (18) and (19) are different forms of the same equation.

## 6.0. Revalidation of Aristotle's assertion

Let small spheres of different masses fall down, then $D_b$ and $D_m$ both become constant, as the same liquid is considered for various observations and the acceleration due to gravity $g$ is constant in all cases. Thus, Eq. (1.19) may be written as

$$v = Z\,M \tag{1.20}$$

where $$Z = \frac{\left(1 - \frac{D_m}{D_b}\right) g}{6\pi\eta r} \tag{1.21}$$

Now the small sphere falls with constant velocity. The constant velocity is equal to average velocity, thus

$$v_{av} = Z\,M \tag{1.22}$$

$$v_{av}\ \alpha M$$

Thus, average velocity is proportional to mass. The heavier the body the more quickly it falls than the lighter body. Here the average velocity means the velocity with which the body falls down. It is nothing else but Aristotle's assertion on falling bodies.

## *6.1. Motion of spheres of different radii (masses)*

Let the motion of spheres of different masses but of the same density be considered in a medium, then $D_m$, $D_b$ and $\eta$ become constant so Eq. (1.17) becomes,

$$6\pi\eta r v = D_b V g - D_m V g \tag{1.17}$$

or $$v = \frac{(D_b - D_m) V g}{6\pi\eta r} \tag{1.23}$$

or $$v = \frac{\left(1 - \frac{D_m}{D_b}\right) V D_b g}{6\pi\eta r} \tag{1.24}$$

or $$v = \frac{\left(1 - \frac{D_m}{D_b}\right) g M}{6\pi\eta r} \tag{1.25}$$

The terms $D_m$, $D_b$, $\eta$, $\pi$ etc. can now be regarded as constant. Thus

$$v = K\frac{M}{r} \tag{1.26}$$

$$\text{where } K = \frac{\left(1-\dfrac{D_m}{D_b}\right)g}{6\pi\eta}$$

Further the relation between the mass and the radius can be established as below

$$V = \frac{D_b}{M} \tag{1.27}$$

For the spherical body

$$\frac{4}{3}\pi r^3 = \frac{D_b}{M} \tag{1.28}$$

$$\text{or} \qquad r = \left(\frac{3D_b}{4\pi M}\right)^{\frac{1}{3}} = \frac{k}{M^{\frac{1}{3}}} \tag{1.29}$$

where $k = \left(\dfrac{3D_b}{4\pi}\right)^{\frac{1}{3}}$

Substituting the value of $r$ from Eq. (1.29) into Eq. (1.26),

$$v = K\,\frac{M^{\frac{4}{3}}}{k} \tag{1.30}$$

$$\text{or} \qquad v = z\ M^{\frac{4}{3}} \tag{1.31}$$

where $z = K / k$.

So, if the body falls with constant velocity, then both constant velocity and average velocity become equal. So Eq. (1.31) becomes,

$$\text{or} \qquad v_{av} \alpha M \tag{1.32P}$$

$$\text{Also} \qquad v_{av} = zM \tag{1.22}$$

Thus, Eq. (1.32P) based upon Stokes law is nothing else but Eq. (1.2) which is the equation for Aristotle's assertion.

***Aristotle's assertion***

Average velocity $\propto$ mass of the falling body

$$v_{av}\ \alpha M$$

or
$$v_{av} = ZM \tag{1.2}$$

***Stokes law***

$$v_{av} \propto M \tag{1.32P}$$

$$v_{av} = zM \tag{1.33}$$

So Aristotle's assertion which has been abandoned after 2000 years of continuous use in fact even now is as useful as Stokes law in explaining the falling bodies. Aristotle's assertion in this regard can be more useful than Stokes law if some sensitive and extended experiments are conducted, as Stokes law is confirmed in very a narrow range [6].

Aristotle has mentioned that puffs of smoke rise upward. But did not give assertion for rising bodies like falling bodies.

## *6.2. Confirmation of Stokes law*

Now it is clear that Stokes law (applicable under five postulates) reduces to Aristotle's assertion. Thus, as Stokes law is valid likewise Aristotle's assertion is valid. Therefore, the abandoned Aristotle assertion is also valid in the region where Stokes law is valid. Thus it is important to analyze the conditions under which Stokes law is valid.

Arnold confirmed Stokes law for small spheres of Rose metal of radii 0.002 cm with an accuracy of a few tenths of 1%. It implies that bodies of radii more than 0.002 cm do not move with constant velocity, i.e. fall with variable velocity [6]. If this variable velocity is greater for the heavier body (even under some conditions) then under those conditions Aristotle's assertion is valid. However, under these conditions Stokes law is not valid.

Such experiments can be conducted in more viscous fluids like glycerine, the coefficient of viscosity for glycerine is 1058.4 times that of water ($\eta$ for glycerine is 1.069 decapoise and $\eta$ for water is 0.00101 decapoise). Thus, in glycerine spheres of radii more than 0.002 cm may fall with constant velocity as viscous force $F = 6\pi\eta rv$

is more in case of glycerine. It can be checked experimentally that the region of validity of Stokes law increases. Therefore, Aristotle's assertion in the case of falling bodies is at least as useful as Stokes law; and if some experiments, as suggested above, are successful then it would be more useful.

Thus, it is evident that had Arnold's data (1910) been available in Galileo's time (1564–1642) then definitely Galileo may have drawn such conclusions. So, Aristotle's assertion would have been correct in this region.

## Frequently asked questions

1. **What is Aristotle's assertion about falling bodies?**
   **Answer:** According to Aristotle (384–322 BC) the heavier bodies have higher tendency to fall than lighter one's.

   **Heavier body falls quickly than lighter body**

   Aristotle maintained that the Earth is the natural home of the material bodies, so the material bodies fall down to have the natural position. This assertion got the immediate support from the fact that a sheet or bit of paper falls more slowly than a metallic ball. Aristotle did not mention anything about falling bodies in liquids, the discussion is only confined to air.

2. **What is the mathematical form of Aristotle's assertion?**
   **Answer:** Aristotle's assertion is a statement that heavier body falls quickly than the lighter body. There is no equation which describes this assertion? However, it can be described in proportionality form as

   **The falling tendency ∝ mass of body**

   Aristotle did not express the falling tendency in terms of any physical quantity. Definitely acceleration was not defined at that time, which was defined by Galileo. The falling tendency may be regarded in terms of average velocity.

   $$v_{av} \propto M$$

   It can be changed to an equation by introducing the proportionality coefficient $Z$ as

   $$v_{av} = ZM$$

This equation is being formed for the first time, neither Aristotle nor any other scientist formulated it. It is further useful in understanding the Aristotle's assertion. For completeness, the resultant weight and resultant acceleration can be considered as the falling tendency.

3. **How it was regarded as true for over 2000 years?**
**Answer:** A theory only survives on the strength of mathematical consistency or experimental truth. Aristotle demonstrated that a straw (lighter body) falls more slowly than a stone (heavier body) in air.

Time taken by straw (<0.1 gm) to reach at bottom in air = 20 s
Time taken by stone (>100 gm) to reach at bottom in air = 5 s

Thus, the heavier body falls more quickly than the lighter body. Similarly, the body with a mass of 50 kg falls more quickly in water than the body with a mass of 5 gm, like a stone falls more quickly in air than paper or straw. Due to these observations Aristotle's assertion is regarded as true for nearly 2000 years. The above observations are justified even now in some cases. It can never be denied as such.

4. **How did Galileo contradict it?**
**Answer:** It is alleged that Galileo dropped a 10 pound shot and a 1 pound shot from the Leaning Tower of Pisa. Both the shots struck the Earth at same time. The observation was taken with naked eye (if ever it was taken). Thus, Galileo preached that all bodies fall equal distances in equal intervals of time.

Time taken by 10 pound shot to reach ground = 10 s
Time taken by 1 pound shot to reach ground = 10 s

Thus, heavier or lighter bodies fall onto the ground at the same time. It is contrary to Aristotle's assertion that the heavier body falls more quickly than the lighter. If Aristotle were true, then the 10 pound shot should have fallen more quickly than the 1 pound shot. Basically in this regard the difference in the mass of the falling bodies is significant. But Galileo did not conduct experiments taking bodies of different shapes in different media. Even experiments from the Leaning Tower of Pisa are doubtful but definitely Galileo drew conclusions from his various other experiments. The Leaning Tower of Pisa has a height of

55.86 m and has 294 steps. This is considered an apocryphal tale, its only source being Galileo's secretary. To draw scientific conclusions, all possible observations are to be taken in account.

5. **Can falling tendency of a body be expressed in terms of resultant weight?**
**Answer:** The weight ($W = mg$) of a body is the force with which the body is attracted by the Earth. The greater the weight the higher the attraction force. The greater the attraction force the more quickly the body falls down towards the Earth.
Thus, weight or resultant weight can be expressed as the falling tendency or Resultant weight $\propto$ Falling tendency

$$\text{Resultant Weight} = (1-\frac{D_m}{D_b})VD_b g = (1-\frac{D_m}{D_b})Mg \qquad (1.15)$$

Resultant weight $\propto M$
or falling tendency $\propto M$

which is Aristotle's assertion about the falling bodies. Thus, in this case Aristotle's assertion is validated if the resultant weight is regarded as the falling tendency.

6. **What is similarity between Stokes law and Aristotle's assertion?**
**Answer:** Both the Stokes law and Aristotle's assertion are mainly meant for falling bodies. Both Aristotle's assertion and Stokes law predict that 'the falling tendency of a body' is proportional to its mass. But the Stokes law is well defined for the falling bodies (in terms of equations) whereas in the case of Aristotle's assertion the conditions are not specified. The Stokes law is quantitative in nature and Aristotle's assertion is qualitative for the falling bodies. The relation between the two is studied for first time.

Terminal velocity on the basis of the Stokes law: Taking various factors in account, we can write

$$v = \frac{\left(1-\frac{D_m}{D_b}\right)gM}{6\pi\eta r} \qquad (1.25)$$

where $D_b$ is the density of the body, $D_m$ is the density of the medium, $V$ is the volume of the body (the volume displaced by the body), $\eta$ is the coefficient of viscosity of the liquid, $r$ is the radius of the sphere and $g$ is the acceleration due to gravity. The above equation can be expressed as

or $$v_{av} = \propto M$$

$$v_{av} = zM \tag{1.32}$$

which is the equation for Aristotle's assertion. Thus, Stokes law revalidates the Aristotle's assertion.

## References


1. Resnick, R. and Halliday, D., *Physics Part I* (Wiley Eastern Limited, New Delhi), pp. 48–49 (Fourty second reprint) (1992).
2. Sharma, A., Speculations in Science and Technology, 20, 1997, 297–300
3. Michel, W.C., *Foundations of Physics* (D. Van Nostrand, London, pp. 255–256 (1968).
4. Niebauer, T.M., Mchugh, M.P. and Faller, J.E., *Phys. Rev. Lett.,* 59, 609–612 (1987).
5. Millikan, R.A., *The Electron* (The University of Chicago Press, Chicago), pp. 93–96 (2nd edition, 10th impression) (1980).
6. Sharma, A., *Acta Ciencia Indica*, P No. 1, pp. 22–24, (1996).

2

# Construction of Water, Glycerine and Ethyl Alcohol Barometers

**First glimpse**

- ✓ Torricelli made a mercury barometer in 1644 and measured pressure as $P = DgH$, but in past 370 years no water barometer has not been formed.
- ✓ The height of the water column in the water barometer must be 10.33 m.
- ✓ Does the viscosity of a liquid (not taken into account till date) has any effect on the height of the liquid in the barometer?
- ✓ Will results of glycerine and water barometers differ from the mercury barometer?
- ✓ The value of acceleration due to gravity $g$ can be measured by this method and used in measurement of the mass of the Earth and other astronomical data.
- ✓ If both methods give different values of $g$, then what would the mass of the Earth be ? It is a big question.
- ✓ If experiments fail then it would lead to reanalysis of the equation $P = DgH$
- ✓ The value of acceleration due to gravity $g$ varies with altitude and depth, it can be compared with $g = P/DH$. Thus the value of $g$ can be cross-checked.
- ✓ The value of acceleration due to gravity varies with altitude and depth, it can be compared with $P = Dgh$. All values must be consistent.

## 1.0. Atmospheric pressure

The envelope of gases surrounding the Earth is called the atmosphere. At the Earth's surface the composition of dry air is 78% nitrogen, 21% oxygen, 0.94% argon, 0.03% carbon dioxide, 0.01% hydrogen, and the remainder includes a small fraction of neon and helium. As we go higher and higher the density of the atmosphere decreases and hence the atmospheric pressure also decreases. As the major part of the oxygen is near the surface of the Earth, so at higher altitudes the we find it difficult to breathe. The 50% of the atmospheric mass is below 5.6 km and 90% of the atmospheric mass is below 16 km. The atmospheric pressure is measured with various barometers. In practical way the value of atmospheric pressure can be understood in the following demonstration.

### *1.1. Glass tumbler experiment*

Take a glass tumbler. Fill it with water up to the brim. Slide carefully a cardboard on its rim, so that no air remains in between the cardboard and the water in the tumbler. Put one hand gently on the cardboard and then invert the glass tumbler.

We find that neither the cardboard nor the water falls down. This is due to the upward thrust acting on the lower surface of the cardboard due to atmosphere pressure. Atmospheric pressure balances the weight of water in the tumbler. Thus atmospheric pressure can be demonstrated by this experiments but cannot measured.

### *1.2. The Magdeburg hemispheres*

The first artificial vacuum was produced in 1643 earlier by Evangelista Torricelli. It had inspired von Guericke to design the world's first vacuum pump. The Magdeburg hemispheres were designed by a German scientist and the mayor of Magdeburg, Otto von Guericke in 1656 demonstrated the air pump which he had invented, and the strength of the atmospheric pressure.

The Magdeburg hemispheres are a pair of large copper hemispheres with mating rims, used to demonstrate the power of atmospheric pressure. When the rims were sealed with grease and the air was pumped out, the sphere contained a vacuum and could not be pulled apart by teams of horses. Von Guericke's demonstration was performed on 8 May 1654 in front of the Reichstag and the Emperor Ferdinand III in Regensburg. Thirty horses, in two teams of 15, could not separate the hemispheres until the valve was opened to equalize the air pressure. In 1656 he repeated the demonstration with 16 horses (2 teams of 8) in his hometown of Magdeburg, where he was mayor.

## 2.0. Some significant terms

### *Fluid*

Fluid is name given to a substance which begins to flow when external force is applied to it.

The examples of the fluid are all liquids and gases – air, water, glycerine, hydrogen etc. Fluids do not have a definite shape and take the shape of the container. The study of the fluid at rest is known as hydrostatics and the study on the fluid in motion is known as hydrodynamics.

### *Pressure in fluids*

In case of solids the pressure can be applied in any direction with respect to the surface. But in case of liquids the pressure must be applied at right angles to the surface of the liquids. It is due to the fact that fluids (liquids and gases) cannot sustain a tangential stress.

The pressure ($P$) is defined as the magnitude of normal force acting on the unit surface area of the fluid. If constant force of magnitude $F$ acts normally to unit surface area $A$, then the pressure acting on the surface is given by

$$P = \text{Normal force/area } A = \frac{F}{A} \tag{2.1}$$

Pressure is a scalar quantity; in the CGS system, the unit of pressure is dyne $cm^{-2}$; in the SI system the unit of pressure is $Nm^{-2}$ or Pascal.

The unit of pressure Pascal is given in honour of the great French scientist and philosopher Blaise Pascal. Another unit of pressure is atmosphere (atm.)

$$1 \text{ atm} = 1.013 \times 10^5 \text{ Nm}^{-2} \text{ (or } Pa\text{)}$$

The unit of pressure used for metrological purposes is called bar

$$1 \text{ bar} = 10^5 \text{ Pa}$$
$$1 \text{ millibar} = 10^{-3} \text{ bar}$$

1 torr is the pressure exerted by a 1 mm mercury column =133.28 Pa

Dimensional formula of pressure is $[ML^{-1} T^{-2}]$

There are many consequences and applications of pressure in daily life. The pressure is inversely proportional to area. For example a sharp needle pierces the skin easily, but a blunt needle does not pierce the skin even if the same force is applied in both cases, as in this case the pressure is lower. Now according to Eq. (2.1) the pressure is higher in case of the sharp needle due to the small area. Hence it pierces the skin easily, as smaller the area the greater is the pressure. Whereas the blunt needle has a larger area and hence less pressure.

### *Thrust*

The total force exerted by a liquid on any surface in contact with it is called the thrust on the liquid.

$$\text{Thrust} = \text{Pressure} \times \text{Area of Surface} \tag{2.2}$$

It has units $N$ in the SI system.

## *2.1. First barometer constructed in 1644*

Italian scientist Evangelista Torricelli (1608–1647) devised a method of measuring the pressure of the atmosphere by the invention of a mercury barometer in 1643. The construction of the mercury barometer can be understood as below.

(i) Take a trough or dish and fill it with mercury.
(ii) Fill a long glass tube with mercury, with one end of the tube is closed and the other one open.
(iii) Close the open end of the tube with the thumb and invert it in the trough or dish filled with mercury.
(iv) The mercury rises in the tube through the height of 76 cm.

Thus, we say that the atmospheric pressure is 76 cm or 0.76 m of mercury at sea. If this observation is taken at higher altitude then the height of the mercury column is less depending upon the height and contents in the atmosphere.

Above the mercury column in the tube there is vacuum known as Torricellian vacuum. Currently there are various types of instruments (barometers) used but our aim is to study the formation of water barometers, glycerine barometers, ethyl alcohol barometers, etc.

### *Typical observations*

Let us assume that two mercury barometers are constructed. In one barometer the diameter of the tube is 10 feet and the other one has a diameter of 1 cm, then the height of the mercury columns in both will be the same, i.e. 76 cm of mercury at sea level. However, in the former case a huge amount of mercury will be used depending upon the diameter of the tube. If the tube filled with mercury is bent or tilted, the height of the mercury column still remains the same. i.e. 76 cm of mercury.

Likewise, in the barometer tube, the height of the mercury column is independent of the length of the tube. If one tube is 20 feet in length, the other one 3 feet then in both the cases the height of the mercury column is 76 cm. Further if the tube is bent or tilted then the height of the mercury column remains the same. Even a capillary tube (as feasible) can be used in the formation of a barometer. Thus, the atmospheric pressure is regarded as 76 cm of mercury in the barometer in all cases at the sea level.

## 3.0. Measurement of pressure

The Italian scientist Evangelista Torricelli (1608–1647) was the first scientist who made serious attempts to measure the atmospheric pressure. In 1643 or 1644 he measured the atmospheric pressure equal to 76 cm of mercury in the barometer at sea level. But at that time there was no equation to calculate the pressure.

It became possible to write mathematical equations to estimate the atmospheric pressure in 1685 when Newton published the law of gravitation in the Principia and defined acceleration due to gravity, $g$. Before this, instruments worked on the basis of experience and best possible logic prevailing at that time. The Pascal law was used under the influence of gravity to derive the equation,

$$P = DgH \tag{2.3}$$

where $D$ is the density of the liquid, $g$ is the acceleration due to gravity and $H$ is the height of the liquid column.

### *3.1. Derivation of the equation of pressure or Pascal's law in presence of gravity*

Consider a liquid of density $D$ contained in a vessel in equilibrium at rest. Imagine a cylinder of liquid in a vessel with cross-sectional area $A$ and length $H$.

The mass of liquid in the imaginary cylinder will be

$$M = \text{volume} \times \text{density} = AH \times D = AHD \quad (2.4)$$

Let $P_1$ and $P_2$ be the pressures of liquid at top and bottom surfaces. The liquid cylinder is under the action of vertical forces

(i) Force $F_1 = P_1A$, acting vertically downwards on the top face of the cylinder.
(ii) Force $F_2 = P_2A$, acting vertically upwards on the lower face of the cylinder.
(iii) Weight, $Mg = AHDg$, vertically acting downwards.

Both upward and downward forces must be equal. As the liquid is in equilibrium at rest, so the vertical cylinder of the liquid will also be in the equilibrium state at rest. Therefore, the net force on it will be zero.

$$F_1 + Mg = F_2$$
$$F_1 + Mg - F_2 = 0$$
$$P_1A + ADHg - P_2A = 0$$
$$P_2 - P_1 = DHg$$
$$\text{Further, } P_2 - P_1 = P$$

Hence we get the desired equation of pressure

$$P = DHg$$

$$H = \frac{P}{Dg}$$

It implies that the pressure exerted by the liquid column depends upon the height of the liquid column and the density of the liquid.

The unit of pressure, Pascal, was defined by simply putting the value of $H$ equal to 0.76 m and density of mercury to 13 600 $kgm^{-3}$

So, Eq. (2.3) becomes

$$P = 13\,600 \text{ kgm}^{-3} \times 0.76 \text{ m} \times 9.8 \text{ ms}^{-2} = 1.013 \times 10^5 \text{ Pa}$$

$$H = \frac{1.013\times10^5}{1000\times9.8} = 0.76 \text{ mm of Hg}$$

This value of pressure can be regarded as standard.

### *3.2 Role of surface tension, angle of contact and viscosity in measurement*

When water barometers, glycerine barometers, ethyl alcohol barometers etc. are fabricated all characteristics of liquids which may influence the results should be taken in account. The tubes other than glass can be considered

**Surface tension:** It is the property of the liquid at rest when behaves like an elastic stretched membrane. The liquid also tries to contract to the minimum possible area. Due to this some insects can move over the surface of water and rain drops are spherical in shape. The unit of surface tension is force per unit length.

**Angle of contact:** There are two types of liquids which wet the container and others that do not wet the container. For example, water wets glass. When glass is dipped in water then its meniscus is concave and the angle of contact is acute i.e. less than 90°. Further, mercury does not wet the walls of the container. When glass is dipped in mercury then its meniscus is convex and the angle of contact obtuse, i.e. more than 90°. So this is the importance of the angle of contact and it plays a significant role in various liquids and tubes used in measurement of pressure. The mercury does not wet the glass, but the water does. So it may affect the height of the water column.

**Viscosity of liquids:** It is the property of liquids which resist s the motion of different layers of liquids. It plays its roles in the tube of the barometer when the liquid tries to come to rest.

According to Newton, the viscous drag or backward dragging force ($F$) between different layers depends

(i) directly proportional to area $A$ of the layer, i.e.

$$F \propto A$$

(ii) directly proportional to the velocity gradient

$$F \propto \frac{dv}{dx}$$

or

$$F \propto A \frac{dv}{dx}$$

The proportionality sign can be removed by introducing the coefficient of proportionality $\eta$

$$F = \eta A \frac{dv}{dx} \tag{2.5}$$

This viscous force acts in the opposite direction.

Units of the coefficient of proportionality:

(i) In the CGS system, the unit of the coefficient of viscosity is

$$\text{dyne } s \text{ cm}^{-2} \text{ or poise}$$
$$1 \text{ Poise} = \text{dyne } s \text{ cm}^{-2}$$

(ii) In the SI system, the unit of coefficient viscosity is

$$\text{N s m}^{-2} \text{ or decapoise}$$
$$1 \text{ decapoise} = \text{Nsm}^{-2}$$
$$\text{Further } 1 \text{ decapoise} = 10 \text{ poise}$$

## 4.0. Why the water barometer has not been constructed since 1644

The first barometer, i.e. mercury barometer, was constructed by the Italian scientist Evangelista Torricelli (1608–1647) in 1643. Why no water barometer has been constructed even after 370 years? Water is the most readily available liquid. Why scientists ignored the construction of the water barometer? If the water barometer is constructed will it lead to abnormal results? Today, the space vehicles fabricated and controlled by scientists, have landed on the surface of the Moon and Mars. The most sophisticated and complicated experiments, involving the Large Hadron Collider, were conducted. So experiments concerning the water barometer are no longer complicated for them. Are there so many questions which can only be answered when the barometer will be constructed? Normally scientists take the established basic laws for granted, but they should be critically analysed. However, critical studies have proved that revisiting fundamental laws is equally useful.

### *What would be the length of the water column in the water barometer?*

In this regard if we use water (1000 kg/m$^3$) instead of mercury in the barometer then the height of the water column must be 10.337 m, i.e.,

$$H(\text{water}) = \frac{P}{D_w g} = \frac{1.013\times10^5\,\text{Pa}}{1000\ \text{kg/m}^3\times9.8\ \text{kg/s}^2} = 10.337\,\text{m} \qquad (2.6)$$

Thus the height of the water column in the water barometer must be 10.337 m which is about 13.6 times more than the height of the mercury column. It may be the real reason that scientists have not constructed the water barometer. It is somewhat complicated to fabricate such a long tube and maintain it. It may be the one reason why the water barometer has not been constructed. If such experiments are unsuccessful then reasons for deviations have to be critically analysed.

French scientist Blaise Pascal (1623–1662) is also believed to have constructed a barometer using red wine and a glass tube 46 feet long (about 12 m). But details on how Pascal estimated the height of the red wine barometer are not available. At that time there was no mathematical equation to estimate the height. In this regard the German experimentalist Otto von Guericke (1602–1686), known for the Magdeburg hemispheres and the vacuum pump, is also believed to have constructed a water barometer. At that time there was no mathematical equation to determine the height of the columns of red wine and water. But details of both experiments are not available. If such equations are confirmed quantitatively then reasons have to be investigated for the deviations. Thus such experiments are very significant.

The height of the liquid column of glycerine and ethyl alcohol barometers can be calculated as in the case of the mercury barometer. Similarly, for glycerine

$$H\,(\text{glycerine}) = \frac{P}{D_w g} = \frac{1.013\times10^5\,\text{Pa}}{1260\ \text{kg/m}^3\times9.8\ \text{kg/s}^2} = 8.204\,\text{m} \tag{2.7}$$

Thus, the height of the glycerine column must be 8.204 m and the length of the tube must be more than 8.204 m.

The density of ethyl alcohol is 785 kg/m$^3$, and the height of ethyl alcohol must be

$$H\ (\text{ethylalcohol}) = \frac{P}{D_w g} = \frac{1.013\times10^5\ \text{Pa}}{785\ \text{kg/m}^3\times9.8\ \text{kg/s}^2} = 13.6\,\text{m} \tag{2.8}$$

Thus, the ethyl alcohol barometer will be the tallest one. The height of tube must be more than 13.6 m

None of the barometer has as yet bee constructed. There is no scientific record in the literature. Hence the formation of such barometer will lead to significant results.

## 5.0. Formation of glycerine, water and ethyl alcohol barometers

These barometers can be constructed exactly in the same way as the mercury barometer. In these barometers, mercury is replaced by other liquids, such as water, glycerine, ethyl alcohol. The only and significant difference in this case is the length of barometer tubes. These lengths are many times more than in the mercury barometer. The filling of the

**Table 1.** Comparison of various characteristics of ethyl alcohol, water, glycerine and mercury

| Characteristic | Ethyl alcohol | Water | Glycerine | Mercury |
|---|---|---|---|---|
| Density ($kg/m^3$) | 785 | 1.000 | 1.260 | 13,600 |
| Coeff. of viscosity (poise) | $1.32 \times 10^{-4}$ | $1.01 \times 10^{-2}$ | 10.69 | $15.5 \times 10^{-3}$ |
| Surface tension (dyne/cm) | 22.3 | 75.6 | 63.1 | 465 |
| Angle of contact | $<90^0$ | $8–9^o$ | – | 137 |
| Capillarity | Rise | Rise | – | Fall |
| Physical behaviour | Wets | Wets | Wets | Does not wet |
| Height of liquid column (m) | 13.16 | 10.33 | 8.202 | 0.76 |

tubes with liquids would be complicated and inversion of the tube in the bucket or drum of liquid would be done with the utmost care. In this case help of the external instrument must be taken. Care would the required to note the reading of the height of the liquid column. The barometer tube must be of sufficient length. The barometer tubes can be made of different type of glass and also of transparent plastic. Then the results must be the same in all types of tubes and liquids. The liquids may wet the glass tube if the angle of contact is acute.

Now let us discuss various possibilities about the experimental results. Until sensitive experiments are conducted, all the possibilities of deviations are equally probable. If the height of the water column in the water barometer is not precisely 10.337 m, then the deviations are confirmed. Also if the height of the glycerine column in the glycerine barometer is different from 8.202 m. If the height of the ethyl alcohol column in the ethyl alcohol barometer differs from 13.6 m. Then deviations are confirmed. The deviations must then be carefully and critically analysed.

(a) It must be noted that the equation

$$P = DHg$$

is derived for a hypothetical or imaginary cylinder in the middle of the liquid in equilibrium with the liquid. But here we are applying a formula which is meant for the imaginary liquid of the cylinder in a practical way. These are simply theoretical assessments; exact results will be obtained experimentally.

(b) Such experiments can be done for various liquids. The shortest column of the liquid would be for the liquid whose density is higher than that of than mercury. The mercury does not wet the walls of the container. Similarly, glycerine is the most viscous liquid; the viscosity of glycerine is 689.67 times that of mercury and 1058 times that of water. It causes the internal resistance of the fluid and retards motion. It may affect the stabilization of the liquid column. The viscous fluids also wet the tube. The pressure may be regarded as the same in all cases.

### *5.1. Two methods to measure acceleration due to gravity g*

The equation of atmospheric pressure involves the acceleration due to gravity $g$. Thus if all other parameters are estimated, then $g$ can be measured from Eq. (2.9). Further, it is well known that acceleration due to gravity decreases with altitude as in Eq. (2.16). It is the established method for measuring $g$. Now the value of $g$ can be measured by both methods and compared. The magnitude of $g$ measured by both methods must be the same. The distance between the Earth and the Sun, the mass of the Earth and other astronomical data are measured with help of $g$. So measurement of $g$ by both methods is extremely important.

#### *Measurement of g with the help of a barometer*

With the help of a barometer we can also find the height of liquid columns at different altitudes. It can be done by measuring acceleration due to gravity $g$ at different altitudes. Practically, the acceleration due to gravity is calculated from both equations. i.e. Eq. (2.9) and Eq. (2.16). We have the equation for the pressure exerted by the liquid

$$P = DHg \tag{2.3}$$

where $D$ is the density of the liquid, $H$ is the height of the liquid column and $g$ is the acceleration due to gravity. If observations are taken at any altitude $h$, then

$$P_h = DH_h g_h$$

or

$$g_h = \frac{P_h}{DH} \tag{2.9}$$

where $g_h$ is the acceleration due to gravity at height $h$; $P_h$ is the atmospheric pressure at height $h$; $H$ is the height of the mercury column in the tube at height $h$; $D$ is the density of the liquid filled at height $h$

### *Variation of g with altitude*

At any height the pressure can be measured with sensitive instruments and the height $H$ of the liquid column with a barometer. In the existing literature, the value of the acceleration due to gravity is not measured with the help of Eq. (2.9), i.e. method of atmospheric pressure.

### *Measurement of g with the help of the law of gravitation*

The other established method for measurement of the acceleration due to gravity is the law of gravitation. According to this law [1], the acceleration due to gravity at any height $h$ is given by

$$g_h = \frac{g}{\left(1+\frac{h}{R}\right)^2} \tag{2.16}$$

Now both values of the equations for the acceleration due to gravity can be compared and should be equal. To assess the value of $g_h$ two factors must be measured

(i) The height of the liquid column in the barometer as required in Eq. (2.9);

(ii) The value of altitude $h$ where $g_h$ is measured as in Eq. (2.16).

The value of $g_h$ has to be the same in the two methods. The acceleration due to gravity is used in measurement of the mass of the Earth and other astronomical data. If different values of $g$ are obtained [1], then it would mean the mass of the Earth has different values!

### *Height of the atmosphere*

The atmospheric pressure is $1.01 \times 10^5$ Pa. Now the height of the atmospheric column, which would exert the same pressure as that exerted by 0.76 m of mercury, can be calculated. However, the atmosphere keeps on decreasing as we move upwards. So the density of the atmosphere is variable. For simplicity, let us assume that the density of the atmosphere is uniform and equal to 1.3 kg/m$^3$, then the height of the liquid column is given by:

The pressure exerted by the atmospheric air column =
the pressure exerted by the mercury column 0.76 m high

$$h \times 1.3 \times 9.8 = 0.76 \times 13600 \times 9.8$$

$$h = \frac{0.76 \times 13\,600}{1.3} = 7951 \text{ m} \sim 8 \text{ km}$$

The following assumptions have been made in the calculating the height of the atmosphere:

(i) The acceleration due to gravity does not change appreciably up to a height of the order of 8 km (height of the typical peak of Mount Everest). According to the equation based upon the law of gravitation ($h$ = 8 km)

$$g_h = \frac{g}{\left(1 + \frac{h}{R}\right)^2}$$

is 9.7755 m $s^{-2}$. The acceleration due to gravity at the surface of the Earth is 9.8 m $s^{-2}$.

(ii) The temperature of the atmosphere is same throughout.

## *5.2. Newton's law of gravitation*

The law of gravitation was stated by Newton in 1687 in the Principia and revolutionised the whole science. But Newton did not discuss the role of the medium between the attracting bodies. Thus, the medium between the attracting bodies is considered to be vacuum or the effect of the medium is insignificant, whereas in Coulomb's law the role of the medium is significant. The force of attraction between any two bodies of the universe is called gravitation or gravitational force.

According to law of gravitation, 'each particle of the universe attracts every other particle. The force of attraction is proportional to the product of masses and inversely proportional to the square of the distance between them':

$$F = G\frac{mM}{r^2} \tag{2.10}$$

where $G$ is the universal gravitational constant. The value of $G$ is experimentally determined and is 6.67 × $10^{-11}$ $Nm^2\,kg^{-2}$. The law of gravitation was stated by Newton in 1687 in the book the Principia and this law revolutionised the science. Newton did not discuss the role of media between bodies attracting each other. The reason may be that this effect is insignificant. However, the measurement of $G$ is taken under stringent conditions. But in Coulomb's law the role of the medium is completely discussed. It is a simple problem and can be easily solved.

### *5.3. Gravitation and weight*

The force with which the Earth attracts a body towards its centre is known as the weight of the body. If $m$ is the mass of the body then its weight on the Earth is given by

$$W = mg \tag{2.11}$$

Equation (2.11) is the analogous form of the second law of motion i.e. $F = ma$, here the acceleration $a$ is replaced by the acceleration due to gravity $g$. It is must be noted that in the Principia Newton did not define or state $F = ma$. According to Newton's second law of motion as stated in the Principia, force is proportional to alteration in velocity. This aspect is discussed in Chapter 8.

Consider a mass $m$ is lying on the surface of the Earth. The gravitational force acting between mass $m$ and the Earth is

$$F = G\,\frac{mM}{R^2} \tag{2.12}$$

where $M$ is the mass of the Earth, $R$ is the radius of the Earth, its value is normally taken 6400 km. Both the weight and the gravitational force of attraction are associated with the body and the Earth.

Since $W$ (weight) = $F$ (force of attraction)

$$mg = G\,\frac{mM}{R^2}$$

$$g = \frac{GM}{R^2} \tag{2.13}$$

which is the expression for the acceleration due to gravity. Thus, the acceleration due to gravity is independent of the mass, shape and size of the body but depends upon the mass and radius of the Earth. This is the reason why the acceleration due to gravity is independent of the mass of the body and falls down with the same acceleration. This deduction holds good in vacuum only as the law of gravitation is silent about the medium in which bodies are placed. The universal constant was also measured in air by Henry Cavendish.

### *5.4. Variation caused by the acceleration due to gravity with altitude (height)*

The acceleration due to gravity is maximum at the surface of the Earth at sea level. It decreases as we move higher; this is an experimental observation. The value of $g$ also decreases as we move deeper into the Earth and becomes zero at the centre of the Earth. This trend should also be observed in Eq. (2.9) and higher and deeper in the Earth. Only then both forces will be consistent with each other.

**Gravity**: The force of attraction exerted by the Earth on the body is called the gravitational pull or gravity

**Acceleration due to gravity:** The acceleration produced in motion of a body under the effect of gravity is called the acceleration due to gravity. It is denoted by $g$ and has the value 9.8 m s$^{-2}$ in SI system. Mathematically it is calculated from the law of gravitation.

It is well known that the acceleration due to gravity on the surface of the Earth is given by

$$g = \frac{GM}{R^2} \tag{2.13}$$

or

$$gR^2 = GM$$

Now consider a body at height $h$ above the surface of the Earth, then the acceleration due to gravity is given by

$$g_h = \frac{GM}{(R+h)^2} \tag{2.14}$$

Substituting $GM = gR^2$ from Eq. (2.13) in Eq. (2.14) gives

$$g_h = \frac{gR^2}{(R+h)^2} \tag{2.15}$$

Now, dividing Eq. (2.15) by Eq. (2.13),

or

$$\frac{g_h}{g} = \frac{R^2}{(R+h)^2}$$

or

$$\frac{g_h}{g} = \frac{R^2}{R^2\left(1+\frac{h}{R}\right)^2}$$

or

$$\frac{g_h}{g} = \frac{1}{\left(1+\frac{h}{R}\right)^2}$$

or

$$g_h = \frac{g}{\left(1+\frac{h}{R}\right)^2} \tag{2.16}$$

It implies that the acceleration due to gravity decreases as we go higher and higher from the surface of the Earth.

(i) At the surface of the Earth $h = 0$, then according to Eq. (2.16) the value of the acceleration due to gravity becomes

$$g_h = g = 9.8 \text{ m/s}^2$$

which is true.

Also, the acceleration due gravity according to Eq. (2.9)

$$g_h = \frac{P_h}{DH} = \frac{1.013\times10^5}{13600\times0.76} = 9.8 \text{ ms}^{-2} \tag{2.17}$$

(ii) Let us theoretically consider that at infinitely large distances, $h = \infty$. Then according to Eq. (2.16)

$$g_h = \frac{g}{\left(1+\frac{\infty}{R}\right)^2} = 0 \tag{2.18}$$

Also under the condition $h = \infty$, $g_h = 0$. At this stage pressure $P$ must be zero, hence

$$g_h = \frac{0}{DH} = 0$$

But the pressure becomes zero at a much lower height; at that point the acceleration due to gravity does not tend to zero.

## 6.0. Comparison of two values of acceleration due to gravity

We have the equation for the pressure exerted by the liquid

$$P = DHg$$

where $D$ is the density of the liquid, $H$ is the height of the liquid column, and $g$ is the acceleration due to gravity. If observations are taken at any altitude $h$, then

$$P_h = DHg_h$$

or

$$g_h = \frac{P_h}{DH} \tag{2.9}$$

where all terms are defined earlier.

At any height the pressure can be measured with sensitive instruments (one is Standard Atmospheric Calculator) and height $H$

of the liquid column with the help of a barometer. The value of the acceleration due to gravity is measured with the help of Eq. (2.9), i.e. the method of atmospheric pressure. The other established method for measurement of the acceleration due to gravity is the law of gravitation. According to the law of gravitation the acceleration due to gravity at any height $h$ is given by

$$g_h = \frac{g}{\left(1+\dfrac{h}{R}\right)^2} \tag{2.16}$$

Now both values of acceleration due to gravity from both equations, i.e. Eq. (2.9) and Eq. (2.16), can be compared and should be equal.

Now let us compare the value of the acceleration due to gravity by both methods, i.e. atmospheric pressure and gravitational attraction at the height of 8 km (the typical height of Mount Everest)

(i) By the method of gravitational attraction. Now Eq. (2.16) becomes

$$g_h = \frac{g}{\left(1+\dfrac{8}{6400}\right)^2} = 9.7755 \text{ m/s}^2$$

Thus the acceleration due to gravity at a height of 8 km is 9.7755 m/s².

**Table II.** Heights of liquid columns of different liquids in tubes of different diameters at different heights

| Sr. No | Height (km) | Pressure (Pa) | Acceleration due to gravity (m/s²) | Mercury $H = P/Dg$ (m) | Water ($H_1$) $H_1 = P/D_1 g$ (m) | Glycerine ($H_2$) $H_2 = P/D_2 g$ (m) |
|---|---|---|---|---|---|---|
| 1 | 0 | 101 300 | 9.8 | 0.76 | 10.33 | 8.202 |
| 2 | 1 | 89 874.57 | 9.7969 | 0.6743 | 9.1737 | 7.2807 |
| 3 | 2 | 79 485.2 | 9.7938 | 0.5967 | 8.1158 | 6.4411 |
| 4 | 3 | 70 108.5 | 9.7908 | 0.5265 | 7.1606 | 5.683 |
| 5 | 4 | 61 640.2 | 9.7877 | 0.4630 | 6.2977 | 4.9981 |
| 6 | 5 | 54 019.9 | 9.7847 | 0.4059 | 5.5208 | 4.3816 |
| 7 | 6 | 47 181 | 9.7816 | 0.3546 | 4.8234 | 3.8281 |
| 8 | 7 | 41 060.74 | 9.7785 | 0.3087 | 4.1990 | 3.3326 |
| 9 | 8 | 35 599.8 | 9.7755 | 0.2677 | 3.6417 | 2.8902 |
| 10 | 25 | 2,511.02 | 9.72388 | 0.01898 | 0.25823 | 0.20494 |
| 11 | 50 | 75.944 | 9.6486 | 0.000578 | 0.00787 | 0.006246 |
| 16 | 60 | 20.314 | 9.6188 | 0.000155 | 0.002111 | 0.001676 |
| 17 | 110 | 0 | 9.4715 | 0 | 0 | 0 |

(ii) By the method of atmospheric pressure. The value of atmospheric pressure at a height of 8 km is 35 599 Pa, as calculated from the standard atmosphere calculator. For convenience the mercury barometer can be considered; the density of mercury is 13 600 kg/m². Then

$$g_h = 35\ 599\ /13\ 600\ H$$

Now experimentally determining the height of the mercury column at an altitude of 8 km, the value of the acceleration due to gravity can be calculated. The height of the mercury column must be such that the acceleration due to gravity from both equations, i.e. Eq. (2.9) and Eq. (2.16), must be the same. Such experiments can be conducted in different ways.

## *6.1. Calculations of heights of various liquid columns*

Now the height of the mercury column can be estimated with the help of the equation for acceleration due to gravity given by

$$g_h = \frac{g}{\left(1+\frac{h}{R}\right)^2} \qquad (2.16)$$

The equation for the atmospheric pressure has the form

$$P = Dg_h H$$

In the following discussion the value of pressure is measured by the standard atmospheric calculator. Thus, measuring $D$ (density of liquid) and $H$ (height of liquid column) the value of $g_h$ can be calculated.

(i) At sea level, the atmospheric pressure is 1 Pa, i.e. 1.013 × $10^5$ $Nm^{-2}$. In all the following cases the value of the acceleration due to gravity is given by the standard atmospheric calculator. The acceleration due to gravity is 9.8 m/s². The value of the acceleration due to gravity is given by

$$g_h = \frac{g}{\left(1+\frac{h}{R}\right)^2} = \frac{g}{\left(1+\frac{0}{R}\right)^2} = g = 9.8$$

Now, substituting this value of the acceleration due to gravity in relevant equations, the heights of liquid columns of various liquids are

Mercury $H_{hg} = \dfrac{P_h}{Dg_h} = 0.76$ m

Water $H_w = \dfrac{P_h}{Dg_h} = 10.33$ m

Glycerine $H_{gly} = \dfrac{P_h}{Dg_h} = 8.202$ m

(ii) At 1 km above the sea level, the atmospheric pressure is 89874.57 $Nm^{-2}$. The acceleration due to gravity is 9.7969 $m/s^2$. The value of the acceleration due to gravity 9.7969 $m/s^2$ is obtained by substituting $h = 1$ km in Eq. (2.16). Now, substituting this value of the acceleration due to gravity in relevant equations, the heights of the columns of various liquids are

Mercury $H_{hg} = \dfrac{P_h}{Dg_h} = 0.6743$ m

Water $H_w = \dfrac{P_h}{Dg_h} = 9.1737$ m

Glycerine $H_{gly} = \dfrac{P_h}{Dg_h} = 7.2807$ m

(iii) At 2 km above sea level, the atmospheric pressure is 79 485.2 $Nm^{-2}$. The acceleration due to gravity 9.7938 $m/s^2$. The value of the acceleration due to gravity is given by Eq. (2.16) Now, substituting this value of the acceleration due to gravity in relevant equations, the heights of the columns of various liquids are

Mercury $H_{hg} = \dfrac{P_h}{Dg_h} = 0.5967$ m

Water $H_w = \dfrac{P_h}{Dg_h} = 8.1158$ m

Glycerine $H_{gly} = \dfrac{P_h}{Dg_h} = 6.4411$ m

The heights of liquid columns in various barometers decrease with altitude, but such significant predictions having far reaching importance are yet to be experimentally confirmed. Even water or glycerine barometers have not been constructed as yet.

(vi) At 3 km above the sea level, the atmospheric pressure is 70 108.5 $Nm^{-2}$. The value of the acceleration due to gravity is given by Eq. (2.16) and is 9.7938 $m/s^2$. Now, substituting this value of the acceleration due to gravity in relevant equation, the heights of the columns of various liquids are

Mercury $$H_{hg} = \frac{P_h}{Dg_h} = 0.5265 \text{ m}$$

Water $$H_w = \frac{P_h}{Dg_h} = 7.1606 \text{ m}$$

Glycerine $$H_{gly} = \frac{P_h}{Dg_h} = 5.683 \text{ m}$$

(v) At 4 km above the sea level, the atmospheric pressure is 61 640.2 $Nm^{-2}$. The value of the acceleration due to gravity is given by Eq. (2.16) and is 9.7877 $m/s^2$. Now substituting this value of the acceleration due to gravity in relevant equations, heights of the columns of various liquids are

Mercury $$H_{hg} = \frac{P_h}{Dg_h} = 0.4630 \text{ m}$$

Water $$H_w = \frac{P_h}{Dg_h} = 6.2977 \text{ m}$$

Glycerine $$H_{gly} = \frac{P_h}{Dg_h} = 4.9981 \text{ m}$$

(vi) At 5 km above the sea level, the atmospheric pressure is 54 019.9 $Nm^{-2}$. The value of the acceleration due to gravity given by Eq. (2.16) is 9.7847 $m/s^2$. Now, substituting this value of the acceleration due to gravity in the relevant equations, the heights of the columns of various liquids are

Mercury $$H_{hg} = \frac{P_h}{Dg_h} = 0.4059 \text{ m}$$

Water $H_w = \dfrac{P_h}{Dg_h} = 5.5208$ m

Glycerine $H_{gly} = \dfrac{P_h}{Dg_h} = 4.3816$ m

The heights of liquid columns in various barometers decrease with altitude, but such significant predictions having far reaching importance are yet to be experimentally confirmed. Water or glycerine barometers have not been constructed as yet.

(vii) At 6 km above the sea level, the atmospheric pressure is 47 181 $Nm^{-2}$. The acceleration due to gravity at this height is 9.7816 $ms^{-2}$. The value of the acceleration due to gravity is given by Eq. (2.16). Substituting this value of the acceleration due to gravity in relevant equations, the heights of the columns of various liquids are

Mercury $H_{hg} = \dfrac{P_h}{Dg_h} = 0.3546$ m

Water $H_w = \dfrac{P_h}{Dg_h} = 4.8234$ m

Glycerine $H_{gly} = \dfrac{P_h}{Dg_h} = 3.8281$ m

(viii) At 7 km above the sea level, the atmospheric pressure is 41 060.74 $Nm^{-2}$. The value of the acceleration due to gravity is given by Eq. (2.16). The acceleration due to gravity at this height is 9.7785 $ms^{-2}$. Now substituting this value of acceleration due to gravity in the equation of heights of the columns of various liquids are

Mercury $H_{hg} = \dfrac{P_h}{Dg_h} = 0.3087$ m

Water $H_w = \dfrac{P_h}{Dg_h} = 4.1990$ m

Glycerine $H_{gly} = \dfrac{P_h}{Dg_h} = 3.3326$ m

(ix) At 8 km above the sea level, the atmospheric pressure is 35 599 Pa. The acceleration due to gravity at this height is 9.7755 ms$^{-2}$. The value of the acceleration due to gravity is given by Eq. (2.16). Substituting this value of the acceleration due to gravity in relevant equations, heights of the columns of various liquids are

$$\text{Mercury} \quad H_{hg} = \frac{P_h}{Dg_h} = 0.2677 \text{ m}$$

$$\text{Water} \quad H_w = \frac{P_h}{Dg_h} = 3.6417 \text{ m}$$

$$\text{Glycerine} \quad H_{gly} = \frac{P_h}{Dg_h} = 2.8902 \text{ m}$$

(x) At 25 km above the sea level, the atmospheric pressure is 2 511.02 Pa. The acceleration due to gravity at this height is 9.72388 ms$^{-2}$, as determined from Eq. (2.16). Now, substituting this value of the acceleration due to gravity in relevant equations, the heights of the columns of various liquids are

$$\text{Mercury} \quad H_{hg} = \frac{P_h}{Dg_h} = 0.01898 \text{ m}$$

$$\text{Water} \quad H_w = \frac{P_h}{Dg_h} = 0.25823 \text{ m}$$

$$\text{Glycerine} \quad H_{gly} = \frac{P_h}{Dg_h} = 0.20494 \text{ m}$$

(xi) At 50 km above the sea level, the atmospheric pressure is 75.944 Pa. The acceleration due to gravity at this height is 9.6486 ms$^{-2}$, as determined from Eq. (2.16). Now substituting this value of the acceleration due to gravity in relevant equations, the heights of the columns of various liquids are

$$\text{Mercury} \quad H_{hg} = \frac{P_h}{Dg_h} = 0.000578 \text{ m}$$

$$\text{Water} \quad H_w = \frac{P_h}{Dg_h} = 0.00787 \text{ m}$$

Glycerine $H_{gly} = \dfrac{P_h}{Dg_h} = 0.006246$ m

(xii) At 60 km above the sea level, the atmospheric pressure is, 20.314 Pa. The acceleration due to gravity at this height is 9.6188 ms$^{-2}$, as determined from Eq. (2.16). Now substituting this value of the acceleration due to gravity, the equation of the heights of the columns of various liquids give

Mercury $H_{hg} = \dfrac{P_h}{Dg_h} = 0.000155$ m

Water $H_w = \dfrac{P_h}{Dg_h} = 0.00211$ m

Glycerine $H_{gly} = \dfrac{P_h}{Dg_h} = 0.001676$ m

The heights of liquid columns in various barometers decrease with altitude, but such significant predictions having far reaching importance are yet to be experimentally confirmed. Even water or glycerine barometers have not been constructed yet. Thus, in some cases the values of *H* are very subtle, hence very sensitive instruments for measurements are needed.

## Frequently asked questions

1. **If a water barometer is constructed, what would be the height of the water column? Similarly, what would be the height of the glycerine column in the glycerine barometer?**

   **Answer:** The height of the water column is given by the equation of pressure $P = DgH$. Thus, the height of the liquid column will be

$$H = \frac{P}{Dg}$$

For the water barometer

$$H\,(\text{water}) = \frac{P}{D_w g} = \frac{1.013 \times 10^5\,\text{Pa}}{1000\,\text{kg/m}^3 \times 9.8\ \text{kg/s}^2} = 10.337\ \text{m}$$

Thus, the height of the water column must be 10.337 m and the length of the barometer tube must be more than this.

For the glycerine barometer

$$H\,(\text{glycerine}) = \frac{P}{D_w g} = \frac{1.013\times10^5\,Pa}{1260\ \text{kg/m}^3\times9.8\ \text{kg/s}^2} = 8.204\ \text{m}$$

Thus, the height of the glycerine barometer must be 8.204 m and the length of the barometer tube must be more than this. The diameter of the tube may be 1 m or 1 mm (depending upon convenience). Further experiments can be conducted in troughs of different sizes, then the amount or magnitude of glycerine or other liquid will vary. Now it can be checked whether the results remain the same or vary.

2. **How will the length of the mercury column vary if the mercury barometer is taken to the Moon?**

**Answer:** The height of the mercury column is given by

$$H = \frac{P}{Dg}$$

The acceleration due to gravity on the Moon is one sixth that of the Earth

$$g_{\text{Moon}} = g/6$$

Further, the atmospheric pressure on the Moon is negligible, it is is far less, $10^{-11}$ torr .

$$P_{\text{Moon}} = 10^{-11}\ \text{torr} = 10^{-11} \times 133.28\ \text{Pa} = 133.28 \times 10^{-11}\ \text{Pa}$$

Thus, the equation for the height of the mercury column on the Moon becomes

$$H = \frac{P}{Dg} = \frac{P_{\text{Moon}}}{Dg_{\text{Moon}}} = \frac{6\times133.28\times10^{-11}}{Dg} \tag{2.19}$$

Further, under the conditions prevailing on the Moon, mercury may not remain in the state as it is on the Earth. Let us assume that a chamber is constructed on the Moon and the pressure is externally maintained equal to one atmosphere (equal to that on the Earth), and the height of liquid columns is calculated. The results will the lead to new findings about measurements of pressure. So if experiments are conducted on the Moon, they will provide the best answer to the question. Such experiments can be conducted at various depths in the sea at different pressures.

3. **In absence of mathematical background, the mercury barometer was fabricated by Torricelli, the red wine barometer was fabricated by Pascal (as quoted in the literature but without details). But when mathematical equations were written after 1685, no such barometers were fabricated.**

**Answer:** The reason is that when a method becomes established then it is not doubted until and unless some compelling evidences are there. The scientific inertia even tries to undo the evidences. Probably this is the reason why Torricelli's experiment was not duplicated for other liquids.

Here an important development took place in 1685 which shook the scientific world, i.e. the law of gravitation was enunciated by Newton. The acceleration due to gravity $g$ was defined. The mathematical equation for pressure was derived, i.e.

$$P = DHg$$

or

$$H = \frac{P}{Dg}$$

It then became feasible to calculate the exact value of the height of the liquid column for various liquids. Thus many such experiments became feasible and it is recommended that these must be conducted for complete understanding of the phenomena.

4 **Is the dependence of the acceleration due to gravity on atmospheric pressure correct?**

**Answer:** The basic equation for $P = DgH$ was derived with the help of $g$ and this perception is widely used. So the method for measurement of pressure was established. The equation for pressure in terms of the acceleration due to gravity was derived mathematically

$$g_h = \frac{P_h}{DH}$$

Under the given assumption the mathematical derivation is correct. Thus, the higher the atmospheric pressure, the higher the value of the acceleration due to gravity. The 50% of atmospheric mass is below 5.6 km and 90% of atmospheric mass below 16 km. Thus, the atmospheric pressure decreases sufficiently at this height. Further 99.99997% of atmospheric mass is below 100 km.

At a height of 25 km, the atmospheric pressure is 2511.02 Pa (pressure at sea level is $1.01 \cdot 10^5$ Pa). Hence the acceleration due to gravity must decrease sufficiently (subject to variation in *H*). If at some height the acceleration due to gravity decreases and becomes zero or tends to zero, then *g* must also approach the same value.

5 **Is it true that equation $P = DHg$ is derived mathematically for an imaginary cylinder in equilibrium with a liquid? How the equation based upon the imaginary cylinder is then used for measurement of pressure?**

**Answer:** Yes. In the derivation of $P = DHg$, an imaginary cylinder of a liquid is considered which is in equilibrium with the rest of the liquid. The areas of the upper and lower surfaces are assumed to be equal in the derivation.

However, when the equation $P = DHg$ is applied for mercury, water, glycerine barometers, etc. then these are practical cases, not imaginary ones. So all possible experiments must be conducted to draw final conclusions.

6 **Why equations $g_h = \dfrac{P_h}{DH}$ and $g_h = \dfrac{g}{\left(1+\dfrac{h}{R}\right)^2}$ are discussed with details?**

The reason is simple, very simple. Both equations give the value of the acceleration due to gravity. Both equations will be consistent if the same and exact value of *g* will be obtained. If both equations give different values of *g*, then the results are inconsistent. Thus, all possible factors have to be analysed. These observations are of paramount importance because the significant factors like the mass of the Earth and astronomical distances are calculkated with the help of *g*.

It is estimated that 99.99997% of atmospheric mass is below 100 km.

Let atmospheric pressure *P* become zero (or tend to zero) at a height of 110 km then

$$g_h = \frac{P_h}{DH} = \frac{0}{DH} = 0$$

**Acceleration due to gravity at altitude:** The value of *g* at a height of 110 km can be calculated from Eq. (2.16)

$$g_h = \frac{g}{\left(1+\frac{h}{R}\right)^2} \tag{2.16}$$

Now $h$ =110 km, $R$ = 6 400 km, then

$$g_h = \frac{g}{\left(1+\frac{110}{6400}\right)^2} = 0.966485154\ g = 9.4715545\ \text{ms}^{-2}$$

Under this condition some peculiar results are obtained. From Eq. (2.9) the value of $g$ must be zero (if at a height of 110 km, the atmospheric pressure is zero, hence $g_h = 0$).

$$g_h = 0$$

Now under similar conditions (at a height of 110 km), the value of the acceleration due to gravity as given by Eq. (2.16) is 9.4715545 ms$^{-2}$.

$$g_h = 9.4715545\ \text{ms}^{-2}.$$

Thus, such experiments are absolutely necessary to understand the dependence of the acceleration due to gravity on the atmospheric pressure.

**Acceleration due to gravity at depth:** The acceleration due to gravity decreases as we go deeper into the Earth. The acceleration due to gravity at depth $d$ is given by

$$g_d = g\left(1-\frac{d}{R}\right) \tag{2.20}$$

At the surface of the Earth $d$ = 0, thus

$$g_d = g\left(1-\frac{0}{R}\right)$$

Thus, the prediction is similar to Eq. (2.16), i.e.

$$g_h = \frac{g}{\left(1+\frac{h}{R}\right)^2} \tag{2.16}$$

$$g_h = \frac{g}{\left(1+\frac{0}{R}\right)^2} = g \tag{2.16}$$

This equation is derived as Eq. (2.16). At the centre of the Earth $d = R$, i.e. depth = radius of the Earth (hypothetical case), the acceleration due to gravity is given by,

$$g_d = g\left(1-\frac{d}{R}\right) = g\left(1-\frac{R}{R}\right) = 0 \tag{2.21}$$

As the acceleration due to gravity has to be same by both the methods, i.e. Eq. (2.9) and Eq. (2.16), then

$$g_h = \frac{P_h}{DH} = \frac{0}{DH} = 0 \tag{2.22}$$

The obvious conclusion is that the atmospheric pressure at the centre of the Earth is zero ($P = 0$) if the values of $D$ and $H$ remains the same. It constitutes speculative discussion only. These experiments can be conducted in specially fabricated chambers where the atmospheric pressure may be changed to desired levels. Then, measuring various parameters, the acceleration due to gravity can be measured. According to Eq. (2.9), the acceleration due to gravity increases. It is again a speculative discussion and encourages that viable experiments cited above must be conducted.

# 3

# Archimedes Principle: The Oldest Established Law

## First glimpse

- ✓ The general public all over the world knew before Archimedes about the observation that a body pulled from water is lighter.
- ✓ According to the anecdote or legend Archimedes enunciated the principle to ascertain the purity of king's crown but there is no documentary proof for this belief. Also, at that time there was a simpler method to determine the density of the crown.
- ✓ At the time of Archimedes, the unit of volume such as one cubic centimeter and one milliliter was not defined. Also, the mass was not defined as we do now. Thus, measurement of density may not have been as accurate as now.
- ✓ It is strange that mathematical equations were formed on the basis of the Archimedes principle (250 BC) 1935 years of its enunciation, i.e. in 1685 when Newton published The Principia and defined the acceleration due to gravity $g$.
- ✓ The Archimedes principle implies that upthrust ($U = VDg$) depends upon the volume of body $V$, the density of medium $D$ and the acceleration due to gravity $g$.
- ✓ The upthrust experienced by the body does not depend upon the shape of the body. Thus, the upthrust ($U = VDg$) for bodies of the same volume is the same irrespective of shape.
- ✓ The effect of the shape of the body on upthrust can be studied experimentally in various cases, i.e. rising, falling and floating bodies.

✓ The magnitude of units of mass and volume have changed time and again in the history of science, so it is one's guess how density is measured precisely? Thus the density of substances varied in past. So which value did Archimedes confirm?.

## 1.0. Archimedes

Archimedes of Syracuse is considered one of the greatest mathematicians in history. In fact, he is believed to be one of the three greatest mathematicians along with Isaac Newton. His greatest contributions to mathematics were in the area of geometry. Archimedes was also an accomplished engineer and an inventor. He discovered how to find the volume of a sphere and determined the exact value of Pi. Archimedes anticipated the integral calculus 2000 years before Newton and von Leibniz.

He was born about 287 BC in Syracuse, Sicily. In the third century BC, Syracuse was a hub of commerce, art and science. At that time Syracuse was an independent Greek city-state with a 500-year history. He was the son of an astronomer and mathematician named Phidias. Some maintain that he belonged to the nobility of Syracuse, and that his family was in some way related to that of Hieron II, King of Syracuse. Archimedes received his formal education in Alexandria, Egypt, which at the time was considered to be the 'intellectual centre' of the world. When he completed his formal studies in Alexandria, he returned and stayed in Syracuse for the rest of his life. As a youth in Syracuse Archimedes developed his natural curiosity and penchant for solving mathematical problems.

According to the legend, King Hieron had commissioned a new royal crown for which he provided solid gold to the goldsmith. When the crown was prepared, King Hieron was suspicious about its purity and goldsmith's honesty. Archimedes was asked to determine whether or not the crown was pure gold or not. Archimedes was perplexed but found inspiration while taking a bath. He noticed that the full bath overflowed when he lowered himself into it. Thus, the tub must be filled to the brim. He then suddenly realized that he could measure the crown's volume by the amount of water it displaced. He knew that since he could measure the crown's volume, all he had to do was discover its weight in order to calculate its density and hence its purity. It is believed Archimedes was so high-spirited about his discovery that he ran down the streets of Syracuse naked shouting, 'Eureka!' which meant 'I've found it!' in Greek. However, there is no documentary proof for this description or non-scientific popular belief [1]. Archimedes uttered the famous saying 'give me a place to stand on and I can move the Earth.'

It implies that the Earth was regarded as stationary, so Archimedes thought of moving it. It is not known whether he ever married or had children. Archimedes died during the Siege of Syracuse when he was killed by a Roman soldier despite orders that he should not be harmed. The Roman soldier who killed him did not know who he was. Archimedes is said to have requested his friends and relatives to place upon his tomb a representation of a cylinder circumscribing a sphere within it.

### *1.1. Origin of Archimedes principle*

According to the T.L. Heath's book published in 1897 in London [1] The Works of Archimedes (On Floating Bodies, Book I) at page no 253, we have Preposition 7 as "A solid heavier than a fluid will, if placed in it, descend to the bottom of the fluid, and he solid will, when weighed in the fluid, be lighter than its true weight by the weight of the fluid displaced."

Thus, the meaning of the principle is clear: 'when the body is weighed in the fluid it will be lighter than its true weight by the weight of the fluid displaced'.

Weight in fluid = True weight – Weight of fluid displaced (decrease in weight).

So in fluids the body becomes lighter than the weight of the fluid displaced.

Nowadays this preposition is mathematically studied as

$$U = \text{decrease in weight} = \text{weight of fluid displaced by the principle} = D_m Vg$$

where $V$ is the volume of the body, $D_m$ is the density of the medium and $g$ is the acceleration due to gravity.

This preposition is very similar to the definition of the Archimedes principle as we understand it now. The same is quoted in other publications as well and also some extensions regarding the Archimedes principle are suggested.

In the simplest way, the Archimedes principle may be defined as

'When a body is wholly or partially immersed in a fluid at

> rest, then it experiences an upthrust (buoyant force) and its weight decreases. The upthrust is equal to the weight of the fluid displaced.'

or

> 'When a body is wholly or partially immersed in a fluid at rest then it experiences an upthrust or buoyant force and its weight decreases, which is equal to the weight of the fluid displaced.'

It is well known that Archimedes is credited with many discoveries in mathematics and engineering. There has been no book written by Archimedes available now in which the Archimedes principle is stated. However, works of Archimedes are available in many publications. Archimedes had stated the principle in original Greek and now it is available in the translated form, as Newton's laws are available as translated from the Latin. Galileo's doctrines are available in the books authored by him. So it is difficult to assess the exact definition of the principle. However, the general theme of the principle is same. However, the exact origin of the definition must be known. This discussion is relevant in the sense that Newton published three laws of motion in the Principia [2]. Out of these, the first and third laws are the same as taught in textbooks all over the world. However, the second law of motion as taught in textbooks was not given by Newton. It is the strangest fact. So the origin of the law must be traced for its proper understanding.

### *1.2. What was the unit of volume and weight at the time of Archimedes?*

What was the unit of volume in days of Archimedes? Now the volume is expressed in terms of milliliter or cubic centimeter. According to available records [3] the volume was defined for the first time in 1889. In 1889, the magnitude of 1 cubic centimeter was regarded as equal to 1 milliliter

$$1 \text{ cc} = 1 \text{ ml}$$

Then in 1907, the value of 1 cubic centimeter was changed and regarded as equal to 0.999972 ml.

$$1\text{cc} = 0.999972 \text{ ml}$$

In 1964, 1 cc was again redefined equal to 1ml

1 cc = 1 ml

So there are considerable variations in the magnitude of the volume even between 1907 and 1964. So which unit of volume, i.e. 1 cc = 1 ml or 1 cc = 0.999972 ml is correct? The difference between two values of 1 cc is $2.28 \times 10^{-3}$. But the Archimedes principle is regarded as true in every situation. It is just possible that the magnitude of 1 cc may change again. Thus, whatever unit of volume Archimedes used was not absolute. If there is a variation in the measurement of volume $2.28 \times 10^{-3}$, then there would be a variation in the measurement of density. The situation is similar with weight which was not defined when Archimedes is believed to have measured the purity of king's crown. Thus, search and research are continuous processes.

The mass equal to kilogram was earlier (before 1799) defined as the mass of one litre of water. But it was replaced by a platinum artifact in 1799 and placed at the French Archive. The litre is the oldest metric unit, defined in 1795. It is the volume of a cube with each side being one tenth of a metre, thus it is equal to 0.001 $m^3$. So it is not clear how Archimedes defined the mass and volume to calculate the density in 250 BC. Also, weight was defined 1937 years after enunciation of the Archimedes principle when Newton enunciated the law of gravitation in the Principia.

In 1795, the unit of mass – gram – was defined in terms of the volume of water, i.e. the mass of one cubic centimeter of water at 4°C or the mass of one litre of water. In 1799 the prototype kilogram was manufactured, on which the current kilogram is based and equal to 1.000025 litres of water. In the past years the International Prototype Kilogram had been found to vary in mass over time (thus variation in density), so the General Conference in Weights and Measures (CGPM) is expected to redefine the mass in terms of Planck's constant in a meeting in 2014. Thus, how Archimedes took the value of mass in his measurements is anybody's guess. Similar is the situation with the volume. Also, no observations are available regarding the measurement of density.

## 2.0. Relevant terms

### *Mass*

**Mass** is a measure of the amount of material in an object. Mass does not change with the body's position, movement or alteration of its shape, unless the material is added or removed (the results of Special Theory of Relativity are not taken in account). The mass is a fundamental property of an object, a numerical measure of its inertia and a fundamental measure of the amount of matter in the object. Mass is measured with a physical balance.

Mass and weight are often misunderstood terms in various branches of physics. The fundamental relation between the mass and the weight is defined by the Second Law and can be expressed as

$$F = ma \tag{3.1}$$

where

$$F = \text{force (N)}$$
$$m = \text{mass (kg)}$$
$$a = \text{acceleration (m/s}^2\text{)}$$
$$a = (v-u)/t. \tag{3.2}$$

$v$ is the terminal velocity, $u$ the initial velocity.

### *Weight*

**Weight** is the gravitational force acting on a body mass. The weight of a body is obtained from the Second law of motion:

$$W = mg \tag{3.3}$$

where $W$ is the weight, measured in Newton ($N$). The gravitational unit of weight is 1 kg.wt (9.8 $N$); $m$ is the mass, measured in kilograms (kg); $g$ is the acceleration due to gravity, it has units (m/s$^2$) and dimensions of acceleration ($M^0LT^{-2}$). Weight is measured with a spring balance.

In the SI system the mass unit is the kg and since the weight is a force – the weight unit is the Newton (N). 1 Newton is a force which produces in a body of mass 1 kg the acceleration equal to 1 m/s$^2$.

A car's mass is 1 000 kg. The weight can be calculated:

$$w = (1000 \text{ kg})(9.807 \text{ m/s}^2) \tag{3.4}$$

$$= 9807 \text{ N}$$
$$= 9.807 \text{ kN}$$

**Fluid:** The fluid is the common name for liquids and gases, e.g. air, water, hydrogen, glycerine, etc. The Archimedes principle is equally applicable to fluids. The magnitude of the upthrust is more in the case of liquids than gases, as the density of liquids is more than that of gases.

**Buoyancy or upthrust:** *Buoyancy* (also known as the *buoyant force*) is the upward force exerted on an object when it is wholly or partly immersed in a fluid. It is a vector having both magnitude and direction. Buoyancy acts upward for the kind of situations encountered in everyday experience. In the SI system the unit of buoyancy is the Newton [*N*]. Upthrust is a force that pushes an object upwards when it is in a fluid (a liquid or gas). Upthrust is also called buoyancy. Due to this the object appear to be lighter in a liquid or weigh less.

**Density:** Density is defined as the object's mass per unit volume.

The density can be expressed as

$$D = \frac{M}{V} \tag{3.5}$$

where

$$D = \text{density} \left(\frac{\text{kg}}{\text{m}^3}\right))$$

$$M = \text{mass (kg)}$$
$$V = \text{volume (m}^3\text{)}$$

The SI units for density are $\left(\frac{\text{kg}}{\text{m}^3}\right)$.

The densities of air, water and steel are 0.001293 kg/m$^3$, 1000 kg/m$^3$ and 7800 kg/m$^3$, respectively. The relative density of substance is defined as 'the ratio of the density of the substance to the density of water at 4°C.'

It can be calculated by just weighing the solid substance in air and water. Mathematically.

**Relative density of solid** = Weight of solid in air/loss in weight in water (3.6)

For a liquid

**Relative density of a liquid** = Loss in weight of liquid/loss in weight of water. (3.7)

**Thrust:** Total force exerted by a liquid on any surface in contact with it, is called the thrust of the liquid. It is denoted by *F;*

**Pressure:** The thrust exerted by a liquid (at rest) per unit area $A$, then pressure is given by $P = F/A$ (3.8)

The unit of pressure in the CGS system is dyne/cm$^2$ and N/m$^2$ in the SI system. 1 N/m$^2$ is also called Pascal (pa)

$$1\ Pa = 10\ \text{dyne/cm}^2$$

The pressure is scalar quality

## 3.0. Measurement of the density of body by the simplest and possibly oldest method

Archimedes may have determined the purity of king's crown by determining its density or relative density. The both measurements are discussed here one by one.

The calculation of density i.e.

$$\text{Density} = \frac{Mass}{Volume} \tag{3.5}$$

requires determination of the volume of the body. According to the anecdote or hearsay (not supported by documentary proofs) while taking bath in the tub, Archimedes found some volume of water overflowed as he entered the tub. The standard treatise of 1897 on the works of the Archimedes principle by T.L. Heath [1] confirmed that there is no proof for the above anecdote. It similar to the legend that Galileo climbed up the Leaning Tower of Pisa and dropped one pound and ten pound shots. This is without any proof. Another one of the series will be discussed, i.e. the second law of motion, as we teach now, is not the same as given by Newton in the Principia.

Volume of water displaced from the tub = volume of body

Now measuring the volume of the water displaced, the volume of the body can be calculated. The body displaces the volume equal to its own volume.

Density = Mass/volume = Mass/volume of water displaced by body

The volume can be measured by immersing the body in a calibrated cylinder and noting the increase in the level of water in the cylinder. It would be the volume of the body. Further, the mass can be measured with the help of a physical balance. Thus the density can be measured by substituting the value of mass and volume in Eq .(3.5). Hence by measuring the density of the crown or body the purity can be checked. Therefore, the purity of the crown can be checked without the help of the Archimedes principle.

**Mass:** The mass is measured with a physical balance.

**Volume:** The volume of the body is equal to the volume of water it displaces. The volume of a regular shape can be measured with mathematical formulae.

## *3.1. Measurement of relative density*

Up to this stage the Archimedes principle has not been used. Thus up to this the mathematical expression for the Archimedes principle, i.e.

$$\text{upthrust or buoyant force} = \text{weight of fluid displaced} = \text{decrease in weight} \tag{3.9}$$

is not required. So at that time the alternate method for measurement of density existed as given by Eqs. (3.5).

However, it must be noted that at that time mathematicians had not distinguished between mass and weight as we do now. Measurements of the density the purity of king's crown (body) can now be estimated.

The purity of the crown can be estimated by calculating the relative density using the Archimedes principle. It can be done by weighing the body in air and water. The relative density can be easily estimated by applying the following equation [4–6]:

$$\text{Relative density of king's crown or body} = \text{Weight of crown in air / loss in weight of crown in water} \tag{3.6}$$

To experimentally determine the density of king's crown we need a spring balance, a beaker filled with water, a plate into which water will overflow. It can be determined in the following steps.

Apparently Archimedes might have performed such steps, if Archimedes had practically determined density by this method. But the experimental observation is not available in the literature.

(i) Tied the crown with a string or body gently.
(ii) Note the reading of the weight of the crown, say it is $W$ in newtons

or

Weight of the crown in air = $W$ newton (3.10)

(ii) Fill the beaker with water.

(iii) Immerse the crown in the beaker filled with water.

(iv) Directly note down the reading of the weight of crown. Let it be $w$ newton

or

Weight of the crown in water = $w$ newton

Now the loss in weight of the crown = $(W-w)$ newton

Loss in weight = (Weight in air –weight in water) (3.11)

The relative density of king's crown = $W / (W-w)$

= Weight in air / (weight in air – weight in water)

The relative density of king's crown = Weight in air/loss in weight (3.12)

Thus, the relative density can be found by weighing the body in air and water. So in this case we need not measure the volume of the crown. Both methods of measurement of the volume are shown in Table I.

So here we have to assume that spring balances or similar instruments existed in days of Archimedes. Thus relative density can be calculated by above method and hence purity of king's crown. By this procedure the relative density of pure gold can be measured and can be compared with the density of the crown.

**Table I.** Comparison of densities of body by the simple method and the Archimedes principle

| Sr. No. | Characteristic | Density by simpler method | Density by Archimedes principle |
|---|---|---|---|
| 1 | Equation | $D$ = mass /volume | Relative density = weight in air / decrease in weight<br>or<br>Relative density = weight in air / $VD_m g$ |
| 2 | Volume | Body displaces water equal to own volume | Measurement of water displaced |
| 3 | Mass | Physical balance | Physical balance |
| 4 | Comments | Archimedes principle is not used | Archimedes principle is used |

So here two straight forth methods existed for measurement of the density of king's crown or body to check the purity, i.e. measuring its density and relative density. Had the crown been studded with gems the measurement of purity would have been difficult. The gems and precious stones have different densities. Hence the anecdote that Archimedes while taking bath in a tub of water invented a method to determine the purity of king's crown is correctly doubted. With this incidence Archimedes only concluded that weight decreases in water. The principle states that upthrust is equal to the weight of the fluid displaced and is quite difficult to calculate comparatively. From this incidence Archimedes concluded that the weight of the body decreases in water. Regional carpenters all over the world constructed boats to cross the river without knowing Archimedes or his principle in various parts of the world.

## 4.0. Derivations of the Archimedes principle or Equation for the Archimedes principle ($U = VDg$) from its definition using the law of gravitation

According to Archimedes principle 'when a body is wholly or partially immersed in a fluid at rest then it experiences upthrust or buoyant force and its weight decreases. The upthrust or decrease in weight is equal to the weight of the fluid displaced.'

The mathematical equations based upon the Archimedes principle became feasible in 1685 after 1935 years of enunciation of principle, when Newton published the Principia [2]. For such long time the principle was regarded as true without mathematical equations. So one can imagine the status of the Archimedes principle in this period. Mathematically, the weight of the body is given by

$$W = \text{Mass} \times \text{acceleration due to gravity} = Mg \qquad (3.3)$$

The mass, density and volume of the body are related with each other as

$$\text{Mass of body} = \text{Density of body} \times \text{Volume of body} = D_b V$$
$$\text{Weight, } W = D_b V g \qquad (3.13)$$

It acts downwards towards the surface of the Earth. As in case of weight, upthrust also involves the acceleration due to gravity (9.8 $ms^{-2}$).

When a body is immersed in a fluid then it experiences upthrust. According to definition of the Archimedes principle, upthrust is given by

$U$ = the weight of the medium (liquid or gas) displaced
= the mass of the liquid displaced × acceleration due to gravity
= the density of medium × volume of body × acceleration due to gravity = $D_m Vg$ (3.14)

While calculating upthrust the body is regarded under static conditions.

According to the definition of the principle: 'when a body is wholly or partially immersed in fluid at rest then it experiences upthrust (buoyant force) and its weight decreases. The upthrust is equal the weight of fluid displaced.'

Upthrust or decrease in weight = weight of fluid displaced by body.

$U$ = decrease in weight = weight of fluid displaced by the principle = $D_m Vg$ (3.14)

The above facts can be experimentally confirmed with sensitive equipment.

## 4.1. Equation for the Archimedes principle ($U = VDg$) from pressure $P = DgH$

While deriving the equation for pressure, the fluid is considered at rest in a vessel. Then in that vessel an imaginary cylinder of liquid in equilibrium with surroundings is considered and pressure $P$ is calculated.

$$P = DgH \tag{3.15}$$

where $D$ is the density of the liquid, $H$ is the height of the column of the liquid and $g$ is the acceleration due to gravity.

In derivation of the Archimedes principle we consider a similar imaginary cylinder of the liquid. Thus finally the equation for upthrust or buoyant force is considered.

Upthrust = buoyant force = weight of fluid displaced = $VDg$ (3.14)

where $V$ is the volume of the body, $D$ is the density of the liquid and $g$ is the acceleration due to gravity.

As in the case of pressure we have to assume an imaginary cylinder of liquid placed in liquid in equilibrium with liquid. Or a body under similar situation can be considered theoretically in the derivation.

Consider a body of mass $M$ completely immersed in a liquid of density $D$ under static conditions. Let $A$ be the area of the cross section of upper and lower surfaces of the body. Thus, the area of both surfaces is precisely equal. Let $h$ be the height of the body. As this body is immersed in the liquid, let $L$ be the depth of the upper face of the body from the free surface of the liquid.

The pressure acting on the upper face and lower faces of the body:

The downward pressure exerted on the upper surface of the body

$$P_1 = LDg \tag{3.16}$$

The upward pressure acting on the lower surface of the body

$$P_2 = (L + h)Dg \tag{3.17}$$

The downward force on the upper surface of the body is

$$F_1 = P_1A = LDgA \tag{3.18}$$

Similarly, the upward force on the lower surface of the body is

$$F_2 = P_2A = (L + h)DgA \tag{3.19}$$

Therefore, the net upward thrust on the body is given by

$$F = F_2 - F_1 = (L + h)DgA - LDgA = AhDg \tag{3.20}$$

But $Ah$ = area × height = $V$ = volume of cylinder

$$\text{Upward thrust} = F = VDg \tag{3.21}$$

Now $V$ (the volume of the liquid displaced by the cylinder) × $D$ (density of liquid) = $M$

= mass of liquid (of volume equal to that of cylinder) displaced

Upward thrust = $Mg$ = mass of liquid (of volume equal to that of cylinder) displaced × $g$ = weight of liquid displaced. (3.22)

The upthrust ($VDg$ or $VD_mg$) may be obtained from its definition or by derivation the body is considered static in the liquid.

## 5.0. Numerical calculations of upthrust

Consider a spherical body of steel having a volume of 100 cm$^3$ is immersed in water of a density of 1 gm per cc. Calculate the upthrust experienced by the spherical body in water.

The upthrust is given by equation

$$U = \text{density of medium} \times \text{volume of body} \times \text{acceleration due to gravity}$$
$$= VD_mg \tag{3.14}$$

The spherical body of the volume 100 cm$^3$ displaces water equal to its own volume. Thus

Volume of water displaced = Volume of sphere = 100 cm$^3$
Density of water = 1 g/cm$^3$
Upthrust = 100 cm$^3$ × 1 gm/cm$^3$ × 980 cm/s$^2$ = 98000 dyne
1 Newton = 10$^5$ dyne
Upthrust = 0.98 N (3.23)

## 6.0. Measurement of density by the method independent of the Archimedes principle

Basically Archimedes may be understood as the perfect scientific brain. Archimedes stated the principle about 2265 years ago. The body displaces the volume of water (fluid) equal to its volume.

$$\text{Volume of body} = \text{Volume of water it displaces} \tag{3.24}$$

The volume of body can be measured with a graduated cylinder, as already mentioned. Equation (3.24) is independent of the Archimedes principle (i.e. decrease in the weight of the body is equal to the upthrust) and is used in the measurement of density.

$$\text{Density} = \text{mass/volume} = \text{mass/volume of water it displaces}$$

Mass can be measured with a physical balance and the 'volume of water displaced' with a graduated cylinder.

Then, determining the purity of the body, the density can be calculated. The same can be also checked by measuring relative density. The purity of the substance can be estimated by these methods. This method is easier than the previous one, as the weight of the body will be taken in air and water and no displaced volume of water is measured. The method of determination of density and relative density may be regarded as independent of the Archimedes principle, as it was well known earlier also that the weight of the body decreases in water. It is an obvious deduction when water is pulled from a well. Further the bigger the body the more volume it displaces, e.g. it can be observed by immersing bodies in a bucket or beaker filled with water.

The phenomena of rising, falling and floating bodies are also discussed on the basis of the Archimedes principle. There is no evidence whether Archimedes himself started using his principle for such phenomena or other scientists did so.

### *6.1. Laws of flotation*

In the Archimedean era there was just one equation,

The weight of body in water = weight of body in air – upthrust
or
The weight of body in water = weight of body in air – decrease in weight (3.25)

But the upthrust and weight were not defined mathematically in the Archimedean era. It is also difficult to predict the conditions of rising, falling and floating bodies. It may have been just possible to understand that heavier bodies fall down, lighter bodies rise, etc. In 1685 Newton defined the acceleration due to gravity *g*, i.e. after 1935 years the equations based upon the Archimedes principle became feasible. Here the equations for weight and upthrust are useful, which were defined after 1935 years of enunciation of the principle. Now the following forces act on the body in a medium.

(i) The weight of the body ($W$), vertically downward through its centre of gravity

$$W = VD_b\, g \tag{3.13}$$

(ii) Upthrust ($U$), acting vertically upwards when the body is immersed in it. Mathematically,

$$\text{upthrust} = \text{weight of fluid displaced} = VD_m g \tag{3.26}$$

The laws of flotation: The law of flotation states that the body

(a) floats if upthrust is equal to the weight of the body,

i.e. weight = upthrust

(b) rises if the upthrust of the body is more than weight

i.e. upthrust > weight

(c) sinks if the weight of the body is more than the upthrust experienced by body.

i.e. weight > upthrust

The weight of the body always acts downward and upthrust upward. Thus, resultant force is the difference of weight and upthrust or buoyant force.

As both forces (i.e. weight and upthrust) compete with each other their net effect or resultant weight

Resultant weight, $W_R$ = weight of body – upthrust or buoyant force

$$= VD_b\, g - VD_m\, g = (D_b - D_m)Vg \qquad (3.27)$$

Various cases of rising, falling and floating bodies will be considered one by one

**(a) A body rises in a medium:** Consider a body of cork is released from the bottom in a tank of water. If the upthrust or buoyant force experienced by the body in the fluid is more than the weight of the body, i.e.

$U > W$

or $VD_m\, g > VD_b g$

or $D_m > D_b$

or $D_b < D_m$

or density of body < density of medium

Thus the body rises upwards in the medium if it is lighter than the medium. For example a cork (240 kg/m$^3$) rises in water (1000 kg/m$^3$). Also in terms of the resultant force,

$$W_R = (D_b - D_m)Vg = -ve$$

$$W_R = (240\text{–}1000)Vg = -760\ Vg \qquad (3.28)$$

The resultant weight is the vector quantity having both magnitude and direction. The negative sign indicates that the body rises upwards. When the body moves upwards then the acceleration due to gravity is also regarded as negative ($-g$).

The density of the medium is more than that of the body, i.e. $D_m > D_b$. Hence,

$$W_R = -ve$$

Hence the body rises upwards in that medium. The distance travelled by the body is given by

$$S = ut + \frac{1}{2}at^2 \tag{3.29}$$

$$v^2 - u^2 = 2as$$

where $S$ is the distance travelled, $u$ is initial velocity, $v$ is terminal velocity and $t$ is time. But none of equations are related with resultant weight. The equations relate distance travelled by the body with acceleration.

**Exception**: Under the influence of external engine, an airplane which has density thousands times more flies or floats in air. It is not consistent with Archimedes principle. Archimedes principle is contradicted (as denser particles float over the lighter medium) in sedimentation and ultracentrifugation experiments as reported by R. Piazza [7]. A striking example of this is rather weird effect that gold particles (spherical in shape) 19.3 gm per cc are seen to float atop lighter surface. It is justified that Archimedes principle is only limiting approximation and generalized Archimedes principle has been given by adding term effective excess buoyancy to upthrust [7]. The similar observations are reported by *D*. de las Heras [8].

**Typical comments**: The shape of rising body is insignificant when it rises upwards in medium.

(b) **Body falls in the medium:** If the weight of the body is more than upthrust or buoyant force experienced by the body in fluid

$$W > U$$

or $\quad VD_b\, g > VD_m g$

or $\quad D_b > D_m$

or density of the body > density of the medium

Thus the body falls downwards in medium if it is heavier than medium or density of the body is more than that of medium. For example a steel ball (7800 kg/m$^3$) rises in water (1000 kg/m$^3$). Also in terms of resultant force,

$$W_R = (D_b - D_m)Vg$$
$$W_R = (7800 - 1000)Vg = 6800Vg \qquad (3.30)$$

The density of the medium is more than that of the body, i.e. $D_m > D_b$. Hence,

$$W_R = +\text{ve}$$

Hence the steel body falls in water freely. The resultant weight is the vector quantity having both magnitude and direction. The positive sign indicates that the body falls downwards. When the body moves upwards then acceleration due to gravity is also regarded as positive ($g$). The resultant weight is not related with distance travelled by the body.

If the body travels in vacuum then $D_m = 0$

$$W_R = D_b Vg \qquad (3.30)$$

The resultant acceleration can be calculated by dividing Eq. (3.30) with mass (6800 V) we get

$$a = g = 9.8\text{m/s}^2$$

Now the distance traveled by the body in time $t$ is given by

$$S = ut + \frac{1}{2}at^2 = \frac{1}{2}gt^2 \qquad (3.31)$$

**Typical comments**: According to mathematical equations, the shape of the falling body is insignificant when the body falls down in the medium (only the density of the body is significant, not other factors). The bodies of steel of the different shapes fall different distances in different times.

(c) **Body floats in the medium:** If the weight of the body is equal to upthrust or buoyant force experienced by the body in fluid.

$$W = U$$
$$\text{Or} \quad VD_b\, g = VD_m g$$
$$\text{or} \quad D_b = D_m$$

A plastic body floats completely submerged in water as its density is equal to that of water. Also, in terms of the resultant force,

$$W_R = (D_m - D_m)Vg = 0$$

or $D_m = D_b$

density of the body = density of medium (3.32)

The Archimedes principle is not used to explain the motion of rising, falling and floating bodies quantitatively. It is not clear whether Archimedes himself extended application of his principle to floating bodies or not. Or it may be possible that by experience people may be knowing that body heavier body (such as stone) falls in air and body lighter body (wood in water) rises upward. People may know that a denser stone falls down in a rarer medium air or water and lighter wood rises in water.

**Typical comments**: For example a steel blade if gently placed on the surface of water floats due to surface tension. The shape of the floating body is insignificant when it floats in the medium. When $D_m = D_b$, the body possesses neither rising nor falling tendency.

A balloon stops rising upward in the atmosphere when body density of body atmosphere becomes equal to that of body balloon. Thus with body help of body Archimedes principle the state of rising, falling and floating bodies can be explained, but qualitatively only. It means body Archimedes principle simply implies that when body density of the body is more than that of body medium, then it falls down. The Archimedes principle does not illustrate

'how much distance' the body falls in 'certain time' in various media.

Similarly, the Archimedes principle implies that when the density of the body is less than that of the medium, then it rises upwards. The Archimedes principle does not illustrate

'how much distance' the body rises in 'certain time' in various media.

Thus Archimedes principle only qualitatively states that when the density of the body is less than that of the medium then it rises upwards. It does not explain the phenomena quantitatively.

## 7.0. Exception to Archimedes principle
## or
## Factors not taken into account by the Archimedes principle

When the body is wholly or partially immersed in a fluid then it experiences the upthrust or buoyant force ($VD_m g$) which competes with weight ($VD_b\ g$). It must be noted that in this regard only the density of the body and that of medium are significance. The other factors include

(i) Shape of immersed body
(ii) Coefficient of viscosity of fluid (gas or liquid)
(iii) Angle of contact
(vi) Surface tension
(v) Any other external force as in the case of airplane in air
(vi) Any type of motion of the medium etc, all are insignificant.

An airplane which has the density thousand times more than that of air floats.

Further a steel blade if gently placed on the surface of water floats due to surface tension.

These factors are not taken in account at all in the Archimedes principle. Only the density of the body and that of the medium are taken in account. In Chapters 4 and 5 the following factors will be considered. For example,

(a) shape of completely submerged floating body
(b) viscosity of medium

are two significant factors which are not taken in account by this principle. Also, in case of rising and falling bodies the Archimedes principle explains the phenomena qualitatively only, not quantitatively. This aspect becomes clear if the distances travelled by heavier and lighter bodies are considered.

## Frequently asked questions

1. **Newton published all his works in the Principia. Is there any book published by Archimedes which illustrates his works?**
   **Answer:** There is no such book written by Archimedes which has survived till date in which the Archimedes principle is stated. However, the works of Archimedes are compiled in many publications. The standard treatise in this regard is The Works of Archimedes, written by

T.L. Heath and published in 1897 in London. Archimedes had stated the principle in original Greek and now it is available in the translated form, as Newton's laws are available in English as translated from the Latin. So it is difficult to assess the exact definition of the principle. However the general theme of the principle is same. But the exact origin of the definition must be known. This discussion is relevant in the sense that Newton published three laws of motion in the Principia. Out of these, the first and third laws are same as taught in textbooks all over the world. However, the second law of motion is taught which is not given by Newton. It is the strangest fact. So the origin of the law must be traced for its proper understanding.

2. **Calculate the distances travelled by spherical and distorted bodies of aluminum (2.7 gm/cc) in water in 1 s?**

**Answer.** In the mathematical equations based upon the Archimedes principle there is no term which may account for the shape of the body. The mathematical equation does not distinguish whether the body is spherical or distorted if density is the same. So according to the principle, the mathematical equations for spherical and distorted bodies of aluminum (different masses) are the same.

The distance travelled is given by the kinematical equation,

$$S = ut + \frac{1}{2}at^2 \tag{3.29}$$

If the body starts from the rest $u = 0$, then

$$S = \frac{1}{2}at^2 \tag{3.33}$$

Now the distance can be calculated if acceleration $a$ is estimated.

$$W_R = (D_b - D_m)Vg = (2.7\text{–}1)980 \text{ dyne} = 1{,}666 \text{ dyne} \tag{3.34}$$

From Eq. (3.34) the acceleration can be calculated as

$$a_r = \frac{W_R}{2.7} = 617.02 \text{ cm/s}^2 \tag{3.35}$$

Then the distance travelled is

$$S = \frac{1}{2}at^2 = \frac{617.02}{2} = 308.5 \text{ cm} \tag{3.36}$$

If the body falls through vacuum, then the equation of the resultant weight becomes

$$W_R = D_b\ Vg = (2.7)980 \text{ dyne} = 2{,}646 \text{ dyne} \quad (3.37)$$

$$a_r = \frac{W_R}{2.7} = 980 \text{ cm/s}^2 \quad (3.38)$$

$$S = \frac{1}{2}at^2 = \frac{980}{2} = 490 \text{ cm/s}^2 \quad (3.39)$$

Theoretically the same distance will be travelled by spherical and distorted bodies of the same density as that of aluminum. Not to speak of experimental verifications, such calculations are not found in the existing literature.

If two bodies of aluminum of the same mass but of different shapes, i.e. sphere and flat plate are thrown in water then the plate falls slowly. However mathematically the equation of the Archimedes principle contains no term which may account for the shape of the body. Similarly, the shape affects motion of other bodies. So this aspect can be experimentally confirmed in sensitive experiments.

In the existing literature there is no evidence whether such experiments were ever conducted. As there are no references then it can be safely concluded that such experiments have not been conducted. So it is the original problem and will add to knowledge if fully explored. Such experiments have far reaching importance as far as fundamental physics (fluid dynamics) is concerned.

3. **Calculate the distances travelled by spherical and distorted bodies of cork (240 gm/cm³) in water in 1 s?**

   **Answer.** Due to lack of experimental data, this case is far more typical than the previous one. It is confirmed that bodies fall with a constant acceleration of 9.8 m/s² in vacuum. But there are no observations in the existing literature as to what is the value of acceleration with which the cork rises in water (as the cork falls in vacuum). The resultant upward acceleration must be different for different bodies in media. Let us calculate the resultant upward force for the body of cork in water

$$W_R = (D_b - D_m)Vg = (240 - 1000)980 \text{ dyne} = -744800 \text{ dyne} \quad (3.40)$$

$$a_r = \frac{W_R}{240} = \frac{744800}{240} = -3103.33 \text{ cm/s}^2 \quad (3.41)$$

$$S = \frac{1}{2}at^2 = \frac{3103.33}{2} = 1551.66 \text{ cm} \quad (3.42)$$

So theoretically the distance travelled by all bodies of different shapes is the same. This prediction has not been checked in experiments yet.

4. **How is the value of 1 cc defined in history?**
**Answer.** The magnitude of 1 cc or 1 cubic centimeter (unit of volume in the CGS system) has undergone different phases [3].
According to the available records the value of 1 cc in 1889, the magnitude of 1 cubic centimeter, was regarded as equal to 1 milliliter

$$1 \text{ cc} = 1 \text{ ml} \quad (3.43)$$

Then in 1907, the value of 1 cubic centimeter was changed and regarded as equal to 0.999972 ml.

$$1 \text{ cc} = 0.999972 \text{ ml} \quad (3.44)$$

In 1964, 1 cc was again redefined as equal to 1 ml

$$1 \text{ cc} = 1 \text{ ml} \quad (3.43)$$

Now it is difficult to understand that how the Archimedes principle was verified:
(i) Between 250 BC–1685, i.e. for 1935 years without mathematical equations and unit of volume.
(ii) Between 1685–1889 i.e. for 204 years between 1685–1889, when there was no definite unit of volume.
(iii) The % age difference between Eq. (3.43) defined in 1889 and Eq. (3.44) defined in 1907 is $2.28 \times 10^{-3}$.
(iv) Thus, there has been an error f $2.28 \times 10^{-3}$% in measurement of volume either between 1907–1964 or after 1964.

The litre is the oldest metric unit, defined in 1795. It is the volume of a cube with each sides of one tenth of a metre, thus it is equal to 0.001 $m^3$. The metric system was developed from 1791 onward. Whatever may be the case, the Archimedes principle is regarded as true in all cases irrespective of the unit of volume. Anyhow, with this background we can critically and mathematically analyze the Archimedes principle in other aspects.

5. **When was the mass defined for first time scientifically and what is the current status?**
**Answer.** In 1795, the unit of mass 'gram' was defined in terms of the volume of water, i.e. mass of one cubic centimeter of water at 4°C or the mass of one litre of water. In 1799, the prototype kilogram was manufactured from which the current kilogram is based and equal to 1.000025 liters of water. In the past years the International Prototype Kilogram had been found to vary in mass over time, so the General Conference in Weights and Measures (CGPM) is expected to redefine the mass in terms of Planck's constant at a meeting in 2014. Thus, how Archimedes took value of mass in his measurements is anybody's guess. Similar is the situation with the volume. Also, no observations are available regarding the measurement of density.

## References

1. Heath, T.L., *The Works of Archimedes principle*, C.J. Clay and Sons, pp. 258–261 (1897).
2. Newton, I., *Mathematical Principles of Natural Philosophy*, pp. 19–20, London, 1727, translated by Andrew Motte from the Latin.
3. Sharma, A. Acta Ciencia Indica, Vol. XXV P, No. 1, 019 (1998).
4. Sharma, A., Speculations in Science and Technology, 20, 1997, 297–300.
5. Green, C.R., *Technical Physics.* Englewood Cliffs: Prentice Hall, 1984, pp. 217–31.
6. Borowitz, S. and Beiser, A., *Essential of Physics.* California: Addison Wesley Publishing Company, 1967, pp. 203–210.
7. Piazza Roberto et al, *Soft Matter*, **8,** 7112–7115 (2012).
8. D. de las Heras et al., Nature Scientific Reports 2, 789, 2012.

4

# The Generalized Form of Archimedes Principle

## First glimpse

- ✓ Archimedes stated the principle in 250 BC, i.e. 2265 years ago. It is the oldest established principle in science.
- ✓ The Archimedes principle has not been confirmed in completely submerged floating balloon experiments in water and other fluids. In these experiments only weight and upthrust act on the balloon.
- ✓ Mathematical equations based upon the principle became feasible after 1935 years, i.e. in 1685 when Newton published the Principia. In the book Newton defined the Law of Gravitation and the acceleration due to gravity $g$.
- ✓ The rising (hot air) balloons were discovered in 1783 by Montgolfier brothers.
- ✓ From mathematical equations it is evident that the mass [as it appears in equations i.e. $m_f = (V + v_f)\ D_w - VD_a$] which the balloon supports is independent of the shape of the balloon.
- ✓ Till date in the history of science no specific experiments have been conducted to check the effect of the shape of completely and independently submerged floating balloons or bodies.
- ✓ Such experiments can be conducted in viscous liquid glycerine and the dense liquid mercury (13 600 kg/m$^3$).
- ✓ The generalized form of the Archimedes principle is $U = fVD_m g$, where $f$ is the coefficient of proportionality.

✓ $f$ takes in account the factors like the shape of body, the viscosity of the medium etc. which are not taken into account by the Archimedes principle.

## 1.0. Introduction

Archimedes (287–212 BC) stated a principle [1] in 250 BC which is known after his name: 'when a body is wholly or partially immersed in fluid at rest then it experiences the upthrust or buoyant force and its weight decreases. The upthrust is equal to the weight of the fluid displaced.'

In 1685, Sir Isaac Newton published the law of gravitation in the Principia and defined the acceleration due to gravity $g$ (9.8 m/s$^2$). The mathematical equations based upon the principle became derivable about 1935 years after the enunciation of the principle. It is very difficult to understand how scientists took the principle for granted for about 1935 years without mathematical backing and treated it as an established law. It is now used as one of the methods to determine the densities of bodies, however, the densities can be measured most accurately by direct methods [2, 3].

According to the anecdote, as mentioned in Chapter 3, Archimedes stated the principle to estimate the purity of king's crown. The purity of any element can be calculated by measuring its density or relative density. To measure density, the volume of the body is required. The volume of the body is the volume which the body displaces i.e.

$$\text{Volume of body} = \text{Mass of water it displaces}$$

The mass can be measured with a physical balance. The volume can be measured easily with a measuring cylinder. Thus, dividing mass with volume density can be estimated.

$$\text{Density} = \frac{\text{Mass}}{\text{Volume}}$$

It is an easier and simpler method and does not require applications of the Archimedes principle strictly as the upthrust is not calculated.

It may be possible that by virtue of experience people may know that a heavier body (such as stone) falls in air and a lighter body (wood in water) rises upward. According to T.L. Heath's book, The Works of Archimedes, published in 1897 in London, in Preposition 7 of Book 1, Archimedes too had stated the same. Even before Archimedes people knew that the mass of the body decreases in fluids. Before this Aristotle (384–322 BC) had maintained that bodies fall down as the earth is their natural place. Smoke rises upward as it has a natural place in the sky. The natural place of planets is in space. According to the Archimedes principle, the body falls in a medium if it is heavier than the fluid, rises if it is lighter than the fluid.

It was also well known to people, even before Archimedes, that a boat can be made of wood which is lighter than water, and air bubbles rise upward in water. The hot air balloons were discovered by Montgolfier brothers in 1782 and the first unmanned flight took place in 1783 which lasted for 10 min. Thus it was amply clear that at that time people knew by common knowledge or experience that lighter bodies rise upward. They did not have any formula for the height of flight of the balloon but they may have estimated it by repeated observations only.

How and when interpretation of rising and falling bodies started in terms of densities? In this regard the reference of Newton's Principia is pertinent. Newton published the law of gravitation and defined acceleration due to gravity $g$. Then the equation of the resultant weight became feasible.

(i) The weight of the body (due to gravity) acts downwards,

$$W = Mg = D_b V g \tag{4.1}$$

(ii) The upthrust (due to the Archimedes principle) acts upwards,

$$U = D_m V g \tag{4.2}$$

Both the forces compete with each other, the net effect is called the resultant weight

$$W_R = (D_b - D_m) V g \tag{4.3}$$

(a) The body rises upwards if the resultant weight is positive or the density of the body > the density of the medium

$$\text{or } D_b > D_m$$

For example a piece of stone falls downward in water.

(b) The body rises upwards if the resultant weight is negative or the density of the body < the density of the medium

$$\text{or } D_b < D_m$$

For example, a piece of wood rises upwards in water.

(c) The body floats if the resultant weight is zero. or the density of the body = the density of the medium

$$\text{or } D_b = D_m \tag{4.4}$$

For example a balloon floats in the atmosphere. The shape of the body is not taken into account. Even then it is a qualitative explanation for the phenomena, not quantitative. In one of the way the floating bodies can be quantitatively studied. The floating bodies may be studied in air, water or in any other media. The associated mathematical equations are based upon the Archimedes principle and Newton's law of gravitation. Even before Archimedes people knew that a piece of wood rises in water and a heavy stone falls down. It may have been known to others as boats and ships were used at that time.

## 2.0. Completely submerged floating balloons

Some experiments regarding floating balloons have not been conducted quantitatively. One of the reasons may be that relevant mathematical equations became derivable only about 1935 years after the enunciation of Archimedes' principle. The mathematical equations have been available since past 327 years but even then no experiments have been conducted. The possible reason may be that in centuries the Archimedes principle got the status of the established law, so it was regarded the principle would be true also in such cases. Many experiments regarding floating bodies may have been conducted in the existing literature but proposed experiments are conceptually different. The established laws must be tested in all possible cases. The following experiments have not as yet been conducted in floating balloons. In independent completely submerged floating balloon experiments only two forces act on the balloon, weight and upthrust.

(i) The weight of the balloon downward
(ii) The upthrust due to the medium or fluid (liquid or gas) in the upward direction

The weight and upthrust naturally balance each other. The balloon is not tied to the spring, physical balance or any other balance with a hook. In existing experiments regarding the measurement of densities the body is tied to the balance from one end. But in floating balloons experiments, the completely submerged balloon is not tied, thus floating independently.

### *2.1. The mathematical equation for the mass which the balloon supports while floating*

Consider a balloon having volume $V$. This volume is filled with air of density $D_a$. The balloon floats in a tank of water supporting (balloon neither rises nor falls) mass $m_f$. The balloon may float just above the bottom of the tank with water, in the middle or upper surface (completely submerged). If adequate weights are attached then the balloon may float just above the bottom of the tank. The balloon is assumed to be floating under perfectly static conditions. Ideally, it must float in the middle of the tank of water. The phenomena can be understood quantitatively by formulating equations taking all factors in account.

(i) The weight of the air-filled balloon consist of the weight of gas filled in it, the weight of the sheath and any additional weight attached to it. Thus

Weight of air filled balloon = mass of balloon × acceleration due to gravity
= [mass of air filled in balloon + mass of sheath + additional mass attached] $g$ (4.5)

Mass of air in balloon = volume of balloon × density of air

$$= VD_a \tag{4.6}$$

$$\text{Mass of sheath of balloon} = m_s$$
$$\text{Mass of additional mass attached} = m_a$$

Now the equation for the weight of the air-filled balloon can be written by substituting various contributions.

$$\text{Weight of air filled balloon} = [VD_a + m_s + m_a]g \tag{4.7}$$

Now writing,

$$m_s + m_a = m_f$$

$m_f$ is the sum of masses of the balloon sheath and the additional mass attached for balancing the balloon.

$$W = (VD_a + m_f)g \tag{4.8}$$

(ii) Applying the Archimedes principle the upthrust experienced by the balloon filled with air of volume $V$ can be calculated. The balloon has the sheath $v_s$ and external mass is attached to the balloon which has volume $v_m$

The upthrust experienced by the air-filled balloon = Total volume of balloon × density of water × acceleration due to gravity = (volume of the air-filled balloon + volume of sheath + volume of additional weights) × density of water × acceleration due to gravity

(4.9)

Volume of air filled in balloon = $V$
Volume of sheath = $v_s$
Volume of additional weights attached = $v_m$
Substituting various terms in Eq. (4.9) we get

$$U = (V + v_s + v_m)D_w g \tag{4.10}$$

Now writing

$$v_s + v_m = v_f \tag{4.11}$$

$$U = (V + v_f)D_w g \tag{4.12}$$

## *2.2. Flotation of a completely submerged balloon or vessel or artifact in water*

Consider a bucket or drum is filled with water. Then air is filled in a balloon or vessel or artifact and properly tightened. The balloon or vessel is then dipped in water and released from the bottom of the tank. Now the weight of he balloon acts in the downward direction and upthrust in the upward direction. Both compete with each other and in magnitudes are given by

$$W = (VD_a + m_f)g \tag{4.8}$$

$$U = (V + v_f)D_w g \tag{4.12}$$

Both weight and upthrust have opposite directions, hence compete with each other. Normally, the upthrust dominates the weight and the balloon or vessel rises upward. Eventually it rests on the upper surface partially submerged in water and air. If additional weights are attached to the balloon then its weight increases and its upward speed decreases. Further if external weights are so judiciously attached that at one stage weight and upthrust become equal, i.e.

Downward weight of vessel = upward upthrust experienced by vessel

$$(VD_a + m_f)g = (V + v_f)D_w g \tag{4.13}$$

where $m_f$ is the mass of the sheath and that of additional weight attached to it. $v_f$ is the volume of the sheath and that of the additional weight attached to it. From Eq. (4.13) the value of $m_f$ can be written as

$$m_f = (V + v_f)D_w - VD_a \tag{4.14}$$

## 2.3. Similar equation in existing literature

Equation (4.14) is used to determine the various variables both theoretically and experimentally. It is used to assess the amount of lighter gases to be filled to balloons so they may rise to a particular height carrying weight. A similar mathematical treatment can be found in Borowitz and Beiser [4], but they have neglected $v_f$ which is very significant. It is being quoted just as a reference, just to understand that this problem is discussed in the existing literature. However all terms are not taken in account. According to the existing literature the weight

Weight of the air-filled balloon = mass of balloon × acceleration due to gravity
= [mass of air filled in balloon + mass of sheath] $g$
= $(VD_a + m_f)g$ (4.8)

The upthrust experienced by the air-filled balloon = Volume of the air-filled in balloon × density of water × acceleration due to gravity

$$U = VD_w g \tag{4.2}$$

Let the balloon is filled with helium gas and it rises upward as upthrust dominates. When the weight and upthrust are balanced then the balloon floats

$$\text{weight} = \text{upthrust}$$
$$(VD_a + m_f)g = VD_w g$$
$$m_f = VD_w - VD_a \tag{4.15}$$

Up to this stage the equation is quoted by Borowitz and Beiser [3] that greater the volume of helium in the balloon, the higher the mass it supports. The term corresponding to the volume of the sheath and additional weights is not taken in account. Further it is not possible to measure the density of air accurately at some height in the atmosphere. So the equations given by Borowitz and Beiser

[4] are not complete. The volume of the balloon sheath can not be neglected. In this case the correct equation is

$$m_f = (V + v_f)\, D_w - VD_a \qquad (4.14)$$

In this case $v_f$ is taken in account which is neglected in Eq. (4.15). So in many respects the study of the topic is incomplete in many respects.

## 3.0. Experimental confirmation of theoretical results

In science scientists firstly enunciate a law based upon rational arguments. Mathematical equations are then formulated on the basis of the proposed law and then checked for consistency. If the mathematical equations are consistent then they are regarded as authentic. The law or mathematical equations are then experimentally confirmed. For example. Newton stated the law of universal gravitation as: 'every body of the universe attracts every other body towards its centre with the force which is proportional to the product of their masses and inversely proportional to the square of the distance between their centers.'

Mathematically,

$$F \propto Mm$$

$$F \propto \frac{1}{r^2}$$

$$F = G\,\frac{mM}{r^2}$$

where $F$ is the force of attraction between two bodies, $m$ and $M$ are the masses of two bodies and $r$ is the distance between their centres. $G$ is the universal gravitational constant and its value is $6.62 \times 10^{-11}$ Nm$^2$ kg$^{-2}$. The law of gravitation is mathematically consistent and used to calculate various astronomical distances and masses of heavenly bodies. Likewise, depending upon the definition of the Archimedes principle, the equation for the mass which the balloon supports is

$$m_f = (V + v_f)\, D_w - VD_a \qquad (4.14)$$

This equation has to be experimentally verified. There is no evidence that this equation has been specifically confirmed experimentally. The mass $m_f$ can be calculated by measuring various parameters: $V$ – the volume of air in the balloon; $v_f$ – the volume of the sheath and that of additional weight if any attached to it; $D_w$ –

the density of water in which the balloon floats; $D_a$ – the density of air with the balloon is filled.

Now the values of various parameters $V$, $v_f$, $D_w$, and $D_a$ can be measured experimentally. Then substituting the values in the equation, the magnitude of $m_f$ can be theoretically calculated. The same value of $m_f$ is experimentally measured and both values of $m_f$ must be the same for shapes if $V$ is the same.

Here the main factor to be noticed is the shape of the balloon. The shape of the balloon does not come in picture at all in the mathematical equations based upon the Archimedes principle. It implies that in case of the completely submerged floating balloon, the shape of the balloon has no role to play. The shape of the balloon may be spherical, flat, elongated, needle- or umbrella-shaped, distorted shape, etc. The value of $m_f$ is theoretically the same. Also in similar experiments (floating bodies) the effect of the coefficient of viscosity can be checked. In the case of floating bodies the Archimedes principle takes into account the density of the body ($D_b$) and that of the medium ($D_m$).

### *3.1. Experimental description*

The mathematical equations are based on the definition of the Archimedes principle. The experimental procedure and results in the initial stage of the experiments are described below. The concrete conclusions can be drawn with sensitive experiments. The trends of the first stage results are qualitative only and expected to act as a catalyst for sensitive experiments. It is just the beginning of the road map but the end would be the results conducted with the most sophisticated experiments. But it is concluded by all experts that such experiments are original [5,6], which is the ultimate reality.

#### *Apparatus*

(i) **Vessels or artifacts or pots:** The first and foremost requirement is vessels or artifacts and pots having a hard shell such that air must be filled in them. They should not elongate or deform when external weights are attached to them. For typical experiments the volume of vessels or pots may have different values, e.g. 50 cm$^3$ to 1000 cm$^3$, or these can be different. The shapes of vessels or artifacts may be spherical, flat, elongated, needle- or umbrella-shaped, distorted shape, etc. The external volume must be carefully measured as it is a significant term.

(ii) **A drum of water:** The vessel or balloon will float in a drum of water, it must be of a sufficient size. The density of water is a significant term and must be accurately measured. Also the experiments can be conducted in different pots (where the balloon floats) to judge the effect of the magnitude of the medium. The same is true for falling and rising bodies.

(iii) **Different weights:** When a balloon is released from the bottom of the tank of water it rises upwards. If a weight is attached to this its rising tendency decreases and it rises slowly. If more weights are attached the rising tendency will further decrease. If the weights are so judiciously attached then at some stage 'the vessel or pot neither rises not falls, just floats'.

Now in this case

(a) The masses of all attached weights are measured.

(b) The volumes of all the attached weights and that of the sheath are measured.

(iv) **A specially fabricated pot of known external volume**: It is difficult to attach various weights to balance the vessel in water so that it may float in water. It would be a tedious process. This problem can be solved by fabricating a pot fitted with a lid so that it may be attached to the vessel. Additional masses can be in the form of fine powder so that the balloon may be balanced properly. Now additional masses will be filled or inserted in this pot purposely opening the lid. They can be attached to the vessel with the help of a hook. It will have two advantages, firstly the mass can be easily measured and, secondly, it would be easier to estimate the volume of the attached weight, to balance the balloon so that it may float.

(v) The other requirements are dishes, etc to collect the displaced water. Graduated cylinders are also needed to measure the displaced water.

## *Procedure*

(i) The balloon or vessel is filled with air or lighter fluid. Then it is left from the bottom and it rises upward.

(ii) Some weights [mass in the mathematical equation (4.14)] are put inside the pot, by removing the pot out of water and then again attached to the vessel.

(iii) Now the upward speed of the balloon or vessel decreases. By adding weights judiciously a stage comes when the balloon or vessel floats, say at the upper surface but completely submerged.

(vi) If at this stage a subtle weight is judiciously attached to the vessel with the help of the pot, then the balloon may float in the middle of the tank of water. Here great precision is required while weights are added.

(v) Further, when the vessel floats in the middle and a subtle weight is attached to it then it may float just above the bottom of tanks. Thus, weights in the form of powder or small granules have to be used purposefully in the pots.

Now the weight which the balloon should support can be calculated by substituting various values ($V$, $v_f$ $D_w$, $D_a$) in Eq. (4.14). Let it be $M_T$, the mass which the balloon supports theoretically.

Now measuring experimentally various values of $V$, $v_f$ $D_w$, $D_a$ etc. the mass can be measured. Let it be $M_E$ (the mass which the balloon experimentally supports), it will be the sum of the mass of the sheath and the mass externally attached (in the pot). If the Archimedes principle holds good then both values must be the same, i.e.

$$M_E = M_T \tag{4.16}$$

Mass experimentally supported by balloon = Mass theoretically predicted.

The equation (4.14) and (4.16) must hold good for balloons of all shapes and for all media.

*First stage observation*: In the first stage experiments, the value of $M_E$ and $M_T$ did not agree.

$$M_E \neq M_T$$

If the balloon is umbrella-shaped, then the balloon or vessel supported more mass than when the balloon was thin pipe-shaped. All the parameters, e.g. $V$, $v_f$, $D_w$ and $D_a$, are same except the shape of the vessel. The qualitative trend of the variation of deviations for different balloons are given in Table I.

**Table I.** The qualitative trend of variation of deviations from the Archimedes principle for air-filled balloons in water, in first stage observations. The sensitive experiments are required for confirmation.

| Sr. No. | Shape variation | Volume | %age deviation |
|---|---|---|---|
| 1 | Spherical to cylindrical | 100–500 $cm^3$ | 2.5–1.25 |
| 2 | Cylindrical to spherical | 500–1000 $cm^3$ | 1.25–3 |

## *Qualitative or expected observations depending upon shape*

Consider the following cases for the sake of simplicity.

(i) If the balloon is spherical and it supports mass $m_1$ gm.

(ii) If the balloon is long pipe-shaped and it supports mass $(m_1-x)$ gm

Such qualitative observations are noted in the first stage experiments. The shape of the body has been observed to play a significant role in the case of rising and falling bodies. So purposeful experiments are required to be conducted to check the effect of the shape of balloons or vessels for floating bodies. The final observations are required with the help of the most sensitive experiments.

## *3.2. Experimental verification of Eq.(4.14)*

To conduct the experiments properly, the value of $v_f$ can be kept constant and determined in the following way. For this, a small pot of a non-hygroscopic material fitted with a hook and a lid should be fabricated. With the help of the hook the small pot can be suspended or attached from a non-hygroscopic balloon or vessel (which also has a hook). To balance the balloon or vessel, the masses can be placed inside the small pot opening the lid carefully outside the water. The additional weights (masses) should be in the form of fine granules or powder because in order to balance the balloon properly, masses of the order of a fraction of milligram may be required. The mass which the balloon supports should be the mass of the pot and the mass of the sheath of the vessel. The rest of the terms can be measured easily. If the external weights (masses) are attached to the balloon one after other then the process would be more cumbersome. The determination of their masses and volumes would be difficult. Thus the use of the pot facilitates the experimental procedures and calculations.

Some first stage experiments were conducted for about 6 months in 1991 for air-filled balloons of different shapes in water. According to the principle $m_f$ depends only on the involved parameters $V$, $v_f$, $D_w$ and $D_a$, and not on its shape. From these first-stage experiments, the dependence of $m_f$ on the shape of the balloon was clearly observed. The volume of balloons varied from almost 100 cm$^3$ to about 500 cm$^3$, the shapes varied from spherical to cylindrical and the deviations ranged from 2.5 to 1.25%. Next when the volume was increased from 500 cm$^3$ to about 1000 cm$^3$ the shape of the balloon varied from cylindrical to nearly spherical and the percentage deviations increased from 1.25 to 2.5%. Here no claim is made for the authenticity of results but one thing is crystal clear that the effect of shape must be quantitatively studied in sensitive experiments with the mass (weight, in general terms) which the balloon supports [7]. Further it must be noted that here the vessel or artifact independently floats under influences of weight and upthrust or buoyant force alone, not attached to the hook of the balance.

In the first stage experiments, the general trend of deviations, i.e. dependence of $m_f$ on the shape of the balloon is observed. The magnitude of deviations was considerably and reasonably higher for balloons of spherical or umbrella shape and lower for those with elongated shapes or pipe-shaped balloons. This magnitude (of deviations) is expected to be higher for umbrella-shaped balloons and lower for long pipe-shaped ones. Apparently, from the first stage experiments it is very clear that the Archimedes principle applied to floating balloons immersed in water under near equilibrium conditions will be true only for a particular shape. However, to confirm the magnitude of the deviations, precise tests with the most sensitive equipment are required. These experiments just indicate the qualitative trends, the shape-dependent deviation in the case of completely submerged floating balloons may be possible. For final conclusions sensitive experiments are required. Here the first trend of deviations is mentioned and accuracy in measurement of parameters is not guaranteed. Such experiments are to be measured precisely and quantitatively. Every new outcome has a humble beginning.

## 4.0. Possible reasons for deviations from Archimedes principle

Some possible deviations from the principle are expected in precise measurements, as the principle does not account for the shape of the body, the effect of depth and size of the medium, etc. Apart from these there are also other subtle factors which affect the results.

(i) **Shape of vessel:** According to Eq. (4.14), $m_f$ depends only upon $V$, $v_f$, $D_w$, and $D_a$ of the balloon and not on its shape. Equation (4.14)

$$m_f = (V + v_f)\, D_w - VD_a \tag{4.14}$$

does not take into account the shape of the body. Here only volumes and densities are significant. No other factor is significant at all, and this aspect can be experimentally checked quantitatively.

Let it be that the internal volume of the balloon or vessel is 1000 cm$^3$ and the volume of the sheath 100 cm$^3$. Now it can have any shape, e.g. spherical, umbrella-shaped, long pipe-shaped or distorted. In all cases the variables i.e. $V$, $v_f$, $D_w$, and $D_a$ relating to the vessel remain the same, hence $m_f$ should be the same. But it is evident from the first stage experiments that $m_f$ also depends upon the shape. So some sensitive experiments are required for confirmation.

(ii) **Effect of shape on the theoretical derivation**

For balloons or vessels of different shapes, the thrusts (pressure × area of the cross-section of the base) on the bases will be different. In the theoretical derivation of the Archimedes principle, an imaginary cylinder of liquid is considered with the areas of upper and lower surfaces being precisely equal

$$F_1 = P_1A \quad = LDgA \tag{4.17}$$

$$F_2 = P_2A \quad = (L+h)DgA \tag{4.18}$$

$$F \ = F_2 - F_1 = (L+h)DgA - LDgA = AhDg = VDg \tag{4.19}$$

If the areas of the upper and lower surfaces are not equal then the upthrust equation is not obtained [8,9]. Thus, this factor (shape of balloon or vessel) must be confirmed experimentally as it plays a significant role in the case of falling and rising bodies.

(iii) **Depth and size of the medium:** Now the vessel may float at any point in the drum of water. If the balloon is floating at the upper surface then if some additional weights are attached to it the position of the vessel is lowered in the drum. Thus it starts floating at increasing depth, say in the middle of the tank. If the added mass is greater then it will float at the bottom.

It means that the balloon may float at the upper surface in the middle of the tank or at the bottom and $m_f$ should remain the same. The balloon will only change its vertical position when the additional mass is added to it, however infinitesimally small. But it may not be true on the basis of the following established fact.

If a submarine is to be lowered to an increasing depth, then more water (equivalent to an additional weight in the case of the balloon) has to be put into the tanks. The submarine has ballast tanks and auxiliary or trim tanks, that can be alternately filled with water or air. When the submarine is on the surface, the ballast tanks are filled with air and the submarine's overall density is less than that of the surrounding water. As the submarine dives, the ballast tanks are flooded with water and the air in the ballast tanks is vented from the submarine until its overall density is greater than the surrounding water and the submarine begins to sink.

So it can be concluded that the balloon may support slightly more mass (weight) at the bottom than at the upper surface of the water. It is observed in the first stage experiments that the effect of depth is not as strong as that of shape in such experiments. Also, according to Eq. (4.14), $m_f$ remains the same irrespective of the fact that the balloon floats in a bucket, drum or tank of water at any position if the density of water remains the same. So some very sensitive experiments are required, as the effect of the size of the medium is observed in the first stage experiments that this effect is theoretically as weak as that of the depth.

According to the principle, $m_f$ is independent of the depth at which the balloon floats in the tank of water. In Eq. (4.14), i.e.

$$m_f = (V+v_f)\, D_w - VD_a \tag{4.14}$$

there is no term which directly accounts for the effect of depth. Consider a tank of water 1 m deep. As the density of water is regarded as same at every depth so the mass which the balloon supports is independent of depth.

Further, to understand the effect of the size of the medium, the experiments can be conducted in a beaker, bucket and tank of water. This effect is not likely to be significant in floating vessel experiments [8, 9].

(vi) **Effect of the coefficient of viscosity**

The body floats if weight and upthrust are equal. Using this condition the mass which the floating balloon supports is given by Eq. (4.14), i.e.

$$m_f = (V + v_f) D_w - VD_a \tag{4.14}$$

Equation (4.14) is completely independent of the coefficient of viscosity of the medium in which the balloon floats. It implies for floating bodies that the coefficient of viscosity is completely irrelevant according to the Archimedes principle. However such experiments in a viscous medium, like glycerine, involving floating bodies, may lead to significant results. The density of the body (thin foil or flat) should be nearly equal to that of glycerine. In that case some significant results are expected. It must be noted that the density of glycerine is 1.26 gm per cc, whereas the coefficient of viscosity of water at 20°C is 0.00101 decapoise and that glycerine 1.069 decapoise. Thus, the coefficient of viscosity is 1058.4 times more than that of water and the density is only 1.2622 times more. The effect of viscosity on floating bodies can be discussed separately for bodies of different shapes [8]. Such effects may be tested for other bodies having different shapes and the density nearly equal to that of the medium.

## *4.1. Effect of density of additional mass attached*

According to Eq. (4.14), if 1000 $cm^3$ of air of density 0.001293 g $cm^{-3}$ is supplied to a vessel it will the float in water with a density of 1 gm $cm^{-3}$ supporting the mass equal to 1.098.707 g if $v_f$ is 100 $cm^3$.

$$m_f = (V + v_f)D_w - VD_a = (1000 + 100) - 1000 \times 0.001239 = 1098.707 \text{ gm} \tag{4.20}$$

If the mass which the vessel of any arbitrary shape supports ($m_f$) is found to be equal to 1099.8057 g or 1097.6083 *g*, then the percentage deviation from the principle will be 0.1. In the case that the vessel is completely evacuated, Eq. (4.20) shows that the value of $m_f$ turns out to be 1100 g:

$$m_f = (V + v_f) D_w = (1000 + 100)\, 1 = 1100 \text{ gm} \tag{4.21}$$

Furthermore, it is also evident that $v_f$ (the volume of the sheath and masses) is quite significant in the calculation of $m_f$. The magnitude of $v_f$ varies with the density of the masses used. For example, 1 kg each of aluminum (2.7 g $cm^{-3}$), mercury (13.6 g $cm^{-3}$) and platinum (21.5 g $cm^{-3}$) have the volumes 370.3703 $cm^3$, 73.529 $cm^3$ and 46.511 $cm^3$ respectively.

Even glycerine which has a density of 1.26 g $cm^{-3}$ can be used as the weight (like solids) but for this the pot to be fabricated must be quite large as 1 kg of glycerine will have the volume of 793.65 $cm^3$. So if additional masses of aluminium, silver and platinum are used for balancing the same vessel then $m_f$ will be different. The value of $v_f$ varies with weights chosen. Now various equations for the mass which the vessel supports can be written as

$$m_f \text{ (aluminium)} = (V+370.3703)\, D_w - VD_a \tag{4.22}$$

$$m_f \text{ (mercury)} = (V+73.529)\, D_w - VD_a \tag{4.23}$$

$$m_f \text{ (platinum)} = (V+46.511)\, D_w - VD_a \tag{4.24}$$

$$m_f \text{ (glycerine)} = (V+793.65)\, D_w - VD_a \tag{4.25}$$

So it is concluded that even if the principle holds good and the additional masses used are of aluminum then it may not hold if the additional masses used are of silver or platinum or glycerine. So, in all cases, experiments have to be conducted separately for balloons of different shapes using different masses (weights); in various fluids of high or low density and viscosity. Thus experimental confirmation of Eq. (4.14) is quite a tedious process and needs to be confirmed in all cases.

## *4.2. Possibility of significant deviations*

According to an anecdote, Archimedes stated the principle for determination of the density of king's crown in 250 BC. However, a simpler method existed at that time, i.e. the volume of the body can be determined by measuring the volume of water (fluid) displaced by it. But mathematical equations based upon the principle became feasible 1935 years after Newton published the Principia. Thus the mathematical equations, including mathematical derivation, became feasible for the Archimedes principle. But these equations were not directly confirmed; however, the principle is regarded as true. This perception is not scientific. This aspect is being contested. The first stage experiments have been conducted and a general trend of the deviations is considered. In some experiments there is a possibility of significant deviations.

**(i) Umbrella-shaped vessel and platinum weights**

In the first stage observations (yet to be finally confirmed in sensitive experiments) an umbrella-shaped vessel supported more weight (mass in mathematical equation) than a long pipe-shaped vessel. Further the mass which the vessel supports depends directly on $v_f$ (volume of the sheath and additional weights, if attached). The value of $v_f$ can be reduced depending upon external weights chosen. The $v_f$ will be smaller if heavier weights are considered. For example platinum is the most dense element (as already stated) for the given mass its volume is lesser. Thus if platinum is used as weight then $v_f$ will have minimum value. Thus, from Eq. (4.24) i.e.

$$m_f \text{ (platinum)} = (V + 46.511)\, D_w - VD_a \qquad (4.24)$$

the mass which the vessel supports will be the smallest, whereas experimentally in the case of the umbrella-shaped balloon the mass which the balloon supports is found to be greater than predicted. If platinum weights are considered, then the mass would be less in Eq. (4.24). Thus, a clear difference between the mathematically predicted and experimentally observed masses is expected. It can be understood in the following way.

(a) The balloon or vessel or artifact must be umbrella- or pipe-shaped. In the first stage observations the umbrella-shaped balloon was found to float by supporting comparatively a larger mass.

(b) The weights attached must be heavy (e.g. platinum) thus according to Eq. (4.14)

$$m_f = (V + v_f)D_w - VD_a \qquad (4.14)$$

If weights are of platinum

$$m_f \text{ (platinum)} = (V+46.511)\, D_w - VD_a \qquad (4.24)$$

The umbrella-shaped balloon is found to support more mass. If the platinum weight is used, then $m_f$ (platinum) is smaller ($v_f$ is smaller).

Hence for this combination (the umbrella-shaped balloon and weights of heavy material) significant deviations are experimentally expected. The reason is that in the first stage observations the umbrella-shaped balloon supported more mass experimentally but according to Eq. (4.24) the predicted mass $m_f$ is comparatively smaller [7]. So, a larger or significant difference in mass (experimental and theoretical) is expected. Thus, a large number of sensitive and sophisticated experiments are required for precise confirmation of the Archimedes principle.

**(ii) Pipe-shaped vessel and glycerine weights**

In the first stage observations (yet to be finally confirmed in sensitive experiments) it was found that the long pipe-shaped vessel supported less weight (mass) than the umbrella-shaped vessel. Further, the mass which the vessel supports depends directly on $v_f$ (the volume of the sheath and additional weights, if any are attached). The value of $v_f$ can be increased depending upon the external weight chosen. $v_f$ (hence $m_f$) will be more if lighter weights like that of glycerine are considered.

For example, glycerine may be one of the lightest weights (as already stated) so for the given mass its volume is greater ($m = DV$). Thus, if glycerine is used as the weight $v_f$ will have the maximum value. Thus, from Eq. (4.25)

$$m_f \text{ (glycerine)} = (V+793.65)\, D_w - VD_a \qquad (4.25)$$

the mass which the vessel supports will be maximum, whereas on the basis of the first stage experimental observations, the pipe-shaped vessel is found to support a lesser mass than predictions. Thus, if the long pipe-shaped vessel is used and glycerine weights are used then some deviations from the Archimedes principle in such cases may be observed. To observe clear deviations, the experiments can be performed in the following way.

(a) The balloon or vessel or artifact must be long, thin, pipe-shaped. In the first stage observations, the long, thin, pipe-shaped balloon was found to float by supporting a lesser mass than predictions.

(b) If the weight attached is of glycerine

$$m_f \text{ (glycerine)} = (V + 793.65)\, D_w - VD_a \qquad (4.25)$$

The mass which the balloon supports would be more theoretically due to higher value of $v_f$. The experimental observation is that a thin pipe-shaped balloon supports less mass. If glycerine weights (heavier than water) are considered then Eq. (4.25) predicts more mass (but experimental mass is expected or theoretical mass is less). Hence for this combination the significant deviations are expected. Therefore, large numbers of sensitive and sophisticated experiments are required for precise confirmation of the Archimedes principle in such experiments.

Such experiments in the identical or modified or extended form can be conducted in the densest liquid mercury (13 600 kg/m$^3$). Mercury has a surface tension of 465 dyne/cm and glycerine 63.1 dyne/cm [7].

## 5.0. The generalized form of the principle

In order to understand and summarize the experimental deviations from Archimedes' principle, one may make a generalization of the principle. The generalization will be valid for bodies of different shapes floating in fluids of different viscosities. It must be understood that the conditions are far from ideal and not strictly hydrostatic. The small scale motion of the fluid and deformation of the vessel must be taken into consideration. Thermodynamical considerations and the possible presence of convection cannot be ignored.

### *5.1. 2264 years old Archimedes principle generalized*

Archimedes (287–212 BC) stated a principle [1] known after his name in 250 BC, 'when a body is wholly or partially immersed in a fluid at rest then it experiences upthrust or buoyant force and its weight decreases. The upthrust is equal to the weight of the fluid displaced.'

$$U = VD_m\, g \tag{4.2}$$

where $U$ is the upthrust or buoyant force, $V$ is the volume of the body, $D_m$ is the density of the medium and $g$ is the acceleration due to gravity. In view of the current consideration and topic described in Chapter 5, the oldest established principle is generalized: 'when a body is wholly or partially immersed in a fluid at rest then it experiences upthrust or buoyant force and its weight decreases. The upthrust is proportional to the weight of the fluid displaced.'

So the definition of the generalized form of the Archimedes principle can be mathematically expressed as

$$U = fVD_m\, g \tag{4.26}$$

where $f$ is the dimensionless coefficient of proportionality which is not equal to unity in all cases, and its value can be determined experimentally. The coefficient of proportionality takes in account the shape of the body, the viscosity of the medium and other subtle factors: thermal effects, surface tension etc. $f$ is like the Reynold number and drag coefficient, both are used in fluid dynamics. The value of $f$ is expected to vary from one situation to other.

### *5.2. Generalized Archimedes principle and the completely submerged floating balloon or vessel or artifact*

In view of the generalized form of the Archimedes principle the mass which the vessel supports can be easily assessed:

$$\text{Weight of air-filled balloon} = (VD_a + m_f)g \tag{4.8}$$

The upthrust experienced by the air-filled balloon = $f$.

The volume of the air-filled balloon × density of water × acceleration due to gravity

$$U = f(V + v_f)D_w g \tag{4.12}$$

Let it be that the helium-filled balloon rises upwards as the upthrust dominates. When the weight and upthrust are balanced the balloon floats.

$$\text{weight of helium-filled balloon} = f$$
$$\text{upthrust experienced by balloon in air}$$
$$(VD_a + m_f)g = f(V + v_f)D_w g$$
$$m_f = f(V + v_f)D_w - VD_a \tag{4.27}$$

Now the speculated experimental results can be explained in the following way.

(i) If the vessel or artifact supports the mass precisely equal to as predicted by the Archimedes principle then the value of $f$ is unity and the original form of the Archimedes principle holds good.

(ii) If the vessel or artifact supports more mass than predicted by original form of the Archimedes principle then the value of

$f$ is more than unity. Thus such observations would confirm the generalized form of the Archimedes principle.

(iii) If the vessel or artifact supports less mass than predicted by the original form of the Archimedes principle then the value of $f$ is less than unity. Therefore, such observations would confirm the generalized form of the Archimedes principle. Such experiments are not described in the existing literature. So this theoretical discussion serves as a platform for noble experiments.

### 6.0. Recent developments

Similar observations were made in very sensitive experiments. The Archimedes principle is contradicted (as denser particles float over the lighter medium) in sedimentation and ultracentrifugation experiments as reported by R. Piazza [10]. A striking example of this is a rather weird effect that gold particles (spherical in shape) 19.3 gm per cc are seen to float atop the lighter surface. It is justified that the Archimedes principle is only a limiting approximation and the generalized Archimedes principle has been obtained by adding the term effective excess buoyancy to upthrust [10]. Similar observations are reported by D. de las Heras [11].

The new form of the Archimedes principle as given by Piazza

= the Archimedes principle + effective excess buoyancy (4.28)

The generalized form of the Archimedes principle

$$= f \text{ the Archimedes principle } = f\, VD_m g \quad (4.29)$$

However, here the principle is generalized in the proportionality form, and $f$ is the coefficient of proportionality and no term is added to it like effective excess buoyancy.

## Frequently asked questions

**1 The Archimedes principle was stated in 250 BC, when did mathematical equations based upon it became feasible?**

In 1685, Sir Isaac Newton enunciated the law of gravitation and defined acceleration due to gravity $g$. It proved to be the turning point in the mathematical basis in various fields of physics. Thus, weight was defined as the product of mass and acceleration due to gravity $g$ as

$$W = mg \quad (4.1)$$

The mathematical equation for upthrust was the written on the basis of definition and mathematical derivation. In both cases the expression for the upthrust also involves the acceleration due to gravity $g$

$$U = DVg \quad (4.2)$$

where $D$ is the density of the body, $V$ is the volume of the body or the volume of the fluid displaced by the body, as the body displaces the fluid equal to its own volume. Then the equation for the resultant weight

$$W_R = (D_b - D_m)Vg = (1 - \frac{D_m}{D_b})\, D_b\, Vg \quad (4.3)$$

and resultant acceleration $a_R = (1 - \frac{D_m}{D_b})g$

and other equations based upon the principle were written. Thus, the equation based upon the Archimedes principle became feasible in 1685, i.e. after 1935 years.

**2 Is the theoretical derivation of the Archimedes principle defined for a body of a particular shape?**

Yes it is. In the theoretical derivation of the Archimedes principle, an imaginary cylinder of liquid is considered in equilibrium with liquid, the areas of upper and lower surfaces are regarded as precisely equal. i.e. $A$. The equations of thrusts are written as

$$F_1 = P_1A = LD_mgA \quad (4.17)$$

$$F_2 = P_2A = (L+h)\, D_mgA \quad (4.18)$$

$$F = F_2 - F_1 = (L + h)\, D_mgA - LDgA = AhD_mg = VD_mg \quad (4.2)$$

Only then the result is

$$U = D_mVg \quad (4.2)$$

If the areas are not precisely equal, then the result is not upthrust, $U = D_mVg$, where all terms are already defined. The Archimedes principle is equally applicable to bodies of distorted shapes, but the above derivation is not applicable for such bodies. So this derivation of the Archimedes principle is only meant for bodies of a particular shape when the areas of the upper and lower surfaces of the body are the same.

**3 What is the equation for mass which the balloon (filled with a lighter medium) supports in medium?**

The weight of the balloon acts downward and the upthrust in the upward direction. The weight of the balloon is given by

$$\text{Weight of the air-filled balloon} = (VD_a + m_f)g \qquad (4.8)$$

$$\text{Upthrust experienced by the balloon } U = (V + v_f)D_w g \qquad (4.12)$$

The balloon floats if weight is equal to upthrust

$$m_f = (V + v_f)D_w - VD_a \qquad (4.14)$$

the various terms are already defined.

In the existing literature there is no evidence that the above equations are experimentally confirmed. Some experiments are suggested to confirm the effect of shape on the floating body. For the first time the experiments are reported by the author highlighting the first stage observations.

**4 Is the effect of shape significant for floating bodies, according to the Archimedes principle?**

No, it is not significant at all.

The condition of flotation is the equality of weight and upthrust.

$$\text{Weight} = \text{Upthrust}$$

$$VD_b g = VD_m g$$

$$D_b = D_m \qquad (4.4)$$

Thus, practically for flotation the density of the body must be equal to that of the medium. Flotation of the balloon is independent of all factors other than $D_m$ and $D_b$. There is no other term which accounts for the shape of the body and other factors. A floating body may have different shapes, e.g. spherical, flat, long pipe-shaped or distorted, etc. According to the Archimedes principle, the shape is completely insignificant. The effect of the shape of the body may lead to interesting results in experiments in dense and viscous liquids.

**5 How does the new form of the Archimedes principle as given by Piazza differ from the one given in this discussion?**

The Archimedes principle is contradicted (as denser particles float over the lighter medium) in sedimentation and ultracentrifugation experiments as reported by R. Piazza [10]. A striking example of this is a rather weird effect that gold particles (spherical in shape) 19.3 gm per cc are seen to float atop of the lighter surface.

The new form of the Archimedes principle (in the simplest form) as given by Piazza is

= the Archimedes principle + effective excess buoyancy(4.28)

The generalized form of the Archimedes principle = $f$ Archimedes principle

$$= f\,VD_mg \tag{4.26}$$

However, here the principle is generalized in the proportionality form, and $f$ is the coefficient of proportionality and no term is added to it like effective excess buoyancy.

According to the Archimedes principle, the condition of flotation is

$$D_m = D_b \tag{4.4}$$

The condition of flotation is that the density of the body must be precisely equal to that of the medium, whereas according to the generalized form of the Archimedes principle, the condition of flotation is

$$D_m = f\,D_b$$

Thus, according to the generalized form of the Archimedes principle the condition of flotation is that even lighter or heavier bodies may float in the medium (depending upon the shape of the floating body, the viscosity of the medium, etc.). So these are different forms of the Archimedes principle and practical application in floating bodies.

**6 What is the main difference between the original and generalized forms of the Archimedes principle?**

**Answer:** It can be understood in the following way.

(i) The generalized form of the Archimedes principle is the extended form of the original form of the principle.

Original form: The upthrust is equal to the weight of the fluid displaced.

$$= VD_m g$$

Generalized form: The upthrust is proportional to weight of the fluid displaced.

$$U_{gen} = f\, VD_m g$$

where $f$ is the coefficient of proportionality. If $f = 1$, then both forms are the same, otherwise the generalized upthrust can be less or more than the upthrust.

(ii) Condition of flotation, weight = upthrust

Original form: $VD_b g = VD_m g$ or $D_b = D_m$

i.e. the body floats in the medium, if the density of the body is equal to that of the medium.

Generalized form: $VD_b g = fVD_m g$ or $D_b = fD_m$

It implies that the body can float if its density is different from that of the medium.

(iii) Original form: It states that in explaining statics or dynamics of bodies, only the density of the body and the density of the medium are required.

Generalized form: It states that in explaining the statics and dynamics of bodies, the density of the medium, the density of the body, the shape of the body, the viscosity of the medium, surface tension, etc. all factors are significant.

Thus, the original form of the principle is a special case of the generalized form. Some sensitive and specific experiments may confirm the validity of the generalized form of the principle.

## References


1. Heath, T.L., *The Works of the Archimedes principle*, C. J. Clay and Sons, pp. 258–261, (1897).
2. Abbott, A.F. (1970), *Ordinary Level Physics,* pp. 112–7 and 138–47, New Delhi: Heinemann Educational Books.
3. Green, C.R. (1984) *Technical Physics,* pp. 217–31. Englewood Cliffs: Prentice Hall.

4. Borowitz, S. and Beiser, A. (1967) *Essential of Physics,* pp. 203–10. California, Addison Wesley Publishing Company.
5. Lindley, D., former Associate Editor Nature, Washington, in report to paper, 1991.
6. Prasad K., former referee Speculations in Science and Technology, in report to paper, 1995.
7. Sharma, A., *Speculations in Science and Technology*, 20, 1997, 297–300, Paper received 25 November 1994: accepted 12 February 1996 (communicated by P. Khastgir)
8. Sharma, A., Acta Ciencia Indica, Vol. XXV, No. 1, 043 (1999).
9. Sharma, A., Acta Ciencia Indica, Vol. XXIII, No. 1, 61 (1997)
10. Piazza Roberto et al, *Soft Matter,* **8** , 7112–7115 (2012)
11. D. de las Heras et al., Nature Scientific Reports 2, 789, 2012.

# 5

# Prediction of Indeterminate Form Of Volume From Archimedes Principle

**First Glimpse**

- ✓ The 2265 years old Archimedes principle is undoubtedly regarded as the oldest established law.
- ✓ At the time of Archimedes mathematical equations involving $g$ did not exist, so it was not completely analysed mathematically by Archimedes.
- ✓ Even concepts of viscosity and surface tension did not exist at the time of Archimedes.
- ✓ Consequently, when the Archimedes principle ($U = DmVg$) is mathematically analysed, Some inconsistent results are obtained. For example, the volume of the body becomes UNDEFINED, i.e $V = \frac{0}{0}$ which is meaningless.
- ✓ Thus, the oldest established principle is generalized for first time, i.e. 'when a body is wholly or partially immersed in a fluid at rest then it experiences upthrust and its weight decreases which is proportional to the weight of the fluid displaced.' Mathematically, $U = f\,DmVg$.
- ✓ When the generalized Archimedes principle, $U = f\,DmV$g, is used then the exact volume is obtained, i.e. $V = V$.
- ✓ Some experiments are required to confirm the generalized form of the Archimedes principle.
- ✓ $f$ is the coefficient of proportionality, it depends upon factors like the shape of the body, the coefficient of viscosity,

etc., which are not taken into account by the Archimedes principle.

## 1.0. Decrease in the weight of the body in a fluid (liquid or gas)

Archimedes is generally considered to be the greatest mathematician of antiquity and one of the greatest of all time [1]. Archimedes is credited with many useful discoveries and inventions. One of his greatest principle, which is known after his name, was used in determination of the density of bodies of regular or irregular shapes [2–6]. This aspect is also discussed in Chapter 3. It is well known to everyone since inception of mankind that when a body is immersed in a fluid, it becomes lighter. However, Archimedes went many steps higher and determined the 'magnitude of decrease in the weight of the body in a fluid.' The principle states: 'when a body is wholly or partially immersed in a fluid at rest then it experiences upthrust and its weight decreases which is equal to the weight of the fluid displaced.'

The word fluid is the common name for liquids and gases. The Archimedes principle is applicable to both, but as gases have a lower density so the magnitude of upthrust is small. It is not certain whether Archimedes stated the principle exactly as above in Greek, but is true that the theme of the principle is the same as stated in above definition. Mathematically above definition can be understood as below.

Decrease in weight of body in fluid = Upthrust experienced by body (5.1)

Upthrust experienced by body = Weight of fluid displaced (5.2)

The body loses the weight equal to the upward push it gets from the fluid depending upon its density.

or Decrease in weight of body in fluid = Weight of fluid displaced (5.3)

So the Archimedes principle can be understood mathematically in any way.

### *1.1. The equations in terms of acceleration due to gravity*

Thus, the decrease in the weight of the fluid is indirectly measured.

Decrease in weight of body in fluid = Weight of fluid displaced
= mass of fluid displaced × $g$
(5.4)

Let observation to be taken in water, then

Decrease in weight of body in water = Weight of water displaced
= mass of water displaced × $g$
= Volume of water displaced × density of water × $g$ (5.4)

So, basically, in the Archimedes principle the volume of the fluid displaced by body is measured and is a significant term. Further, in the simplest way

Volume of fluid displaced = volume of body (5.6)

Thus, in the Archimedes principle the volume of the body is a significant factor. The volume of the body (say, 100 cc) may assume different shapes. Archimedes calculated the density of the body by dividing the weight of the body (*mg*) by the decrease in the weight of the body in water, i.e.

Density of solid = Weight of solid in air/decrease in weight in water (5.7)

Also, to determine the density of a body, the volume of the body of irregular shape must be determined. The volume of any body is equal to the volume of water it displaces when immersed in it. It can be easily measured with a graduated cylinder. This fact is independent of the Archimedes principle. Thus, measuring the volume of water displaced, the volume of a body (may be of irregular shape) can be determined. Now, dividing the mass of the body by its volume, the density can be determined; hence the purity of metal can be checked. It is a simpler way:

Density = Mass of body/volume of fluid

The situation would be different if a gold ornament is studded with gems.

## 2.0. General conditions of flotation

It is interesting to note that Archimedes' principle was stated in 250 BC. At that time the concept of weight was not understood as we do now. Newton [7] enunciated the law of gravitation in 1685 and defined acceleration due to gravity, $g$. The modern mathematical equations about weight, upthrust and resultant weight, as we study them now, became feasible 1935 years after the enunciation of the Archimedes principle. Thus, equations for weight ($mg$) and buoyant force ($VD_m g$) became feasible i.e.

$$\text{Weight} = mg \tag{5.8}$$

where $m$ is mass, $g$ is the acceleration due to gravity.

$$\begin{aligned}\text{Upthrust} &= \text{Weight of fluid displaced}\\ &= \text{Volume of medium displaced} \times \text{density of}\\ &\quad \text{medium} \times g = VD_m g \end{aligned} \tag{5.2}$$

where $V$ is the volume of the body, $D_m$ is the density of the medium and $g$ is the acceleration due to gravity.

For a considerable time, Archimedes' principle was understood without genuine mathematical equations. Consequently, even Archimedes had not mentioned $mg$ and $VD_m g$ and simply used terms like true weight and the weight of the fluid displaced. The reason is that these terms were coined after 1935 years when Newton propounded the law of gravitation and defined the acceleration due to gravity $g$. In view of the equations based upon acceleration due to gravity, some new facts come into picture and lead to noble finding including generalization of the principle.

Later, the applications of Archimedes' principle were extended to floating bodies [8]. Definitely the extension of applications took place after 1685. The principle implies that a body floats when the buoyant force is equal to the weight of the body.

$$\text{Weight of body} = VD_b g \tag{5.9}$$

where $D_b$ is the density of the body,

$$\begin{aligned}\text{Buoyant force (upthrust)} &= \text{Weight of fluid (medium) displaced}\\ &= VD_m g\end{aligned} \tag{5.2}$$

where $D_m$ is the density of the medium.

When the body is immersed in a fluid its weight decreases and the net weight is known as resultant weight,

$$\text{Resultant weight (weight of body in fluid)} = D_b Vg - D_m Vg \tag{5.10}$$

The body floats in a medium if

$$\text{Weight of body} = \text{Buoyant force} \tag{5.11}$$

This condition of flotation of bodies will be analyzed in case of completely submerged bodies (balloons or vessels) in fluids. Further, ideally the principle is applicable under static conditions of the body and the medium. If this condition is not obeyed then the external force affects the state of body. Originally, the principle was not stated for determining the conditions of flotation of the bodies, but applications of the principle were extended in such cases. In floating bodies the weight is balanced by the upthrust, whereas in other experiments the body is tied with a string to the hook of the balance. But the situation is different in the case of floating bodies. In the case of completely submerged floating bodies (balloons or vessels) interesting results are obtained under some conditions (different shapes). These deductions serve as a theoretical basis of generalization of the principle and the generalization can be finally confirmed in logical experiments.

## 2.1. Equations for completely submerged floating balloons

The Archimedes principle is applicable in both cases when under some conditions (different shapes), body is wholly or partially immersed in a liquid. Normally, for quantitative analysis the completely submerged body is considered, it facilitates the exact calculation of volume. So in the following sub-sections, the completely submerged bodies are considered. The Archimedes principle is critically analyzed in this way for first time in the past 2265 years.

It is believed that initially the Archimedes principle [2–6] was originally stated to determine the densities of bodies, and later on its applications were extended to rising, falling and floating bodies. Such equations in the case of completely submerged floating balloons, vessels or artefacts, lead to inconsistent results in some cases as described below. Such results are not discussed in the literature yet. It should be noted that the equations which are discussed below became feasible 1935 years after enunciation of the Archimedes principle and led to noble deductions. Obviously, Archimedes was unaware of results.

## 2.2. Air-filled balloon floating in water

Consider a balloon, a vessel or an artefact having internal volume $V$. Let this volume be filled with metals, cork and air. Let $v$ be the mass of the sheath, it also includes the volume of additional mass, $m$, if any is attached to it. Let the density of the medium inside the vessel be $D_m$. This volume floats completely submerged in water, under the influence of weight and upthrust. Both forces acting on the balloon or vessel can be determined as,

Weight of balloon or vessel = Weight of medium filled in vessel + weight of sheath and additional mass (if any) attached to it = mass of medium filled in vessel × $g$ + mass of sheath × $g$

or $$\text{Weight of balloon} = VD_m\, g + mg \tag{5.12}$$

According to the Archimedes principle, the upthrust experienced by the balloon is equal to the weight of the fluid displaced [2].

Upthrust experienced by balloon in water =

weight of water displaced by balloon =
= mass of water displaced × $g$
= volume of water displaced × density of water × $g$
= (internal volume of balloon + volume of sheath) × $D_w$ × $g$

$$= (V + v)\, D_w\, g \tag{5.13}$$

The weight of the body acts downwards and the upthrust in the upward direction. When both forces balance each other, then the balloon floats, i.e.

Downward weight = Upward weight

$$V D_m\, g + mg = (V + v)\, D_w\, g$$

or $$m = (V + v)\, D_w - V D_m \tag{5.14}$$

where $m$ is the mass of the sheath and any other mass if any is attached to it.

Equation (5.14) involves five variables; if the values of four variables are given then the value of the fifth variable can be determined. It is the definition of the equation. Thus, from Eq. (5.14) different values of $V$, $D_w$, $D_m$ and $v$ can be written as

$$v = \left\{\frac{m + V\left(D_m - D_w\right)}{D_w}\right\} \tag{5.15}$$

$$Dw = \frac{\left(m + VD_m\right)}{\left(V + v\right)} \tag{5.16}$$

and $$D_m = \left\{\frac{D_w\left(V + v\right) - m}{V}\right\} \tag{5.17}$$

$$V = \frac{(m - vD_w)}{(D_w - D_m)} \tag{5.18}$$

## 2.3. Calculations of variables, v, $D_m$, $D_w$ and V based upon the original form of the Archimedes principle

The various values of the volume of the sheath, the density of the medium, the density of water and the volume of the medium in the balloon can be calculated from Eqs. (5.15)–(5.18). Here one condition is applied when the density of the medium filling the balloon is equal to that of water (in which the balloon or vessel floats). Mathematically,

Density of medium filled in balloon
= Density of water in which vessel floats (5.19)

$$D_m = D_b \tag{5.19}$$

Now various parameters or variables in the equation will be calculated. If the equation consists of three variables and two are known, then the third can be estimated. It is the basic principle in all or many equations.

(i) Now firstly let us calculate the volume of the sheath, i.e. $v$ in which volume $V$ of the medium is filled. As already stated, $v$ also includes the volume of external mass, if attached to the balloon or vessel. $m$ is the mass corresponding to volume $v$. Further calculations are done under the condition that the density of the medium (substance) filling the balloon is equal to that of water, i.e.

$$D_m = D_w \tag{5.19}$$

Now substituting the value of $D_m$ from Eq. (5.19) into Eq. (5.14), we get

$$\begin{aligned} m &= (V + v)\, D_w - V\, D_m \\ &= (V + v)\, D_w - VD_w = vD_w \end{aligned} \tag{5.20}$$

To calculate the volume $v$ we substitute Eq. (5.20) and Eq. (5.19) into Eq. (5.15)

$$v = \left\{ \frac{(V+v)D_w - VD_w + V(D_w - D_w)}{D_w} \right\} = \frac{vD_w}{D_w} = v \tag{5.21}$$

which is true.

**Numerical value of *v*:** The value of $v$ (volume of the sheath and mass if any is attached to it) is given by

$$v = \left\{\frac{m + V\left(D_m - D_w\right)}{D_w}\right\} \tag{5.15}$$

It can be numerically calculated by substituting values, i.e. $V$, $m$, $D_m$, $D_w$ in Eq. (5.15). Let us assume that the value of volume $V$ is 200 cc, $m = vD_w$ =20 gm, $D_m = D_w$ =1 gm per cc. Then the value of $v$ can be calculated by putting various values in Eq. (5.15)

$$v = \frac{\left(20 - 200[1-1]\right)}{1} = 20 \text{ cc} \tag{5.22}$$

which is true.

(ii) The density of the medium filling the balloon or vessel is given by Eq. (5.16). Let us calculate the value of the density of the medium filling the balloon under the condition when $Dm = Dw$. Now substituting Eq. (5.19) and Eq. (5.20) in Eq. (5.16)

$$D_w = \frac{\left(m + VD_m\right)}{\left(V + v\right)} \tag{5.16}$$

$$D_w = \frac{\left(vD_w + VD_w\right)}{\left(V + v\right)} = \frac{\left(v + V\right)D_w}{\left(V + v\right)} = D_w \tag{5.23}$$

The above deduction is justified as the density of the medium the density of the medium filling the balloon is equal to that of water.

**Numerical calculations of *v*:** As already mentioned, the volume of the medium inside the balloon $V$ is 200 cc, the volume of the sheath is 20 cc, $D_m = D_w$. Equation (5.16) becomes

$$D_w = \frac{\left(m + VD_m\right)}{\left(V + v\right)} = \frac{\left(20 + 200 \times 1\right)}{220} = 1 \text{ gm/cm}^3 \tag{5.24}$$

The above deduction is true the density of the medium filling the balloon or vessel is 1 gm per cc.

(iii) The density of water filling the balloon or vessel is given by Eq. (5.17). Now for completeness of mathematical analysis the density of water can be calculated. Substituting Eq. (5.19) and Eq. (5.20) in Eq. (5.17)

$$D_m = \left\{\frac{D_w(V+v)-m}{V}\right\} \tag{5.17}$$

or
$$D_m = \left\{\frac{D_w(V+v)-vD_w}{V}\right\} = D_w \tag{5.25}$$

which is true: the density of water is equal to that of the medium. Thus, both densities are equal in this case.

**Numerical calculations:** Like $D_m$ as in Eq. (5.16) or Eq. (5.25) the value of $D_w$ can be numerically calculated by substituting various numerical values in the equation. As already mentioned, the volume of the medium inside the balloon $V$ is 200 cc, the volume of the sheath is 20 cc, $D_m = D_w$

or
$$D_m = \left\{\frac{D_w(V+v)-vD_w}{V}\right\}$$

$$= \frac{(200+20)-20}{200} = 1 \text{ gm per cc} \tag{5.26}$$

which is the correct result. The density of water and that of the medium filling the balloon is the same.

(vi) **Indeterminate form of volume**

Let internal volume $V$ of the balloon or vessel be 200 cc filled with platinum (21.5 gm per cc) of volume 2 cc, silver (10.5 gm per cc) of volume 2 cc, cork (0.24 gm per cc) of volume 10 cc and the rest is air and metal. The density of the medium is equal to that of water in which the balloon floats. Let the volume of the sheath be 20 cc and made of a non-hygroscopic material and floats in water. The sheath may be non-hygroscopic if its composition is 93.8% in weight Nylon 12 and 6.2% Santicizer [14].

The completely submerged balloon floats as the density of the medium filling the balloon is equal to that of water in which it floats. Also, $m$ is the mass of volume $v$. Now, under these conditions, volume $V$ can be calculated. The volume of the sheath of the balloon and that of the additional mass attached to the balloon is $v$. The balloon floats as the density of the fluid (substance) inside the balloon, a vessel or an artefact is equal to that of water ($D_m = D_w$). Obviously the volume should turn out equal to $V$, which is the actual value. Equation (5.18) gives the value of volume $V$.

$$V = \frac{(m - vD_w)}{(D_w - D_m)} \tag{5.18}$$

Now substituting Eq. (5.19) and Eq. (5.20), i.e. $D_m = D_w$ in Eq. (5.18)

$$V = \frac{((V + v)D_w - D_w V - vD_w)}{(D_w - D_w)} = \frac{(vD_w - vD_w)}{(D_w - D_w)} = \frac{0}{0} \tag{5.27}$$

or Volume of platinum, silver, cork, metal and air $= \dfrac{0}{0}$

which is the indeterminate form, i.e. the volume of the medium filling the balloon (platinum, silver, cork, metal and gases) becomes undefined but in actual experimental set up the volume is $V$ consisting of metal, wood and air. Thus, the right hand side (RHS) of Eq. (5.27) becomes devoid of units and dimensions which is not defined. This result is meaningless. The dimensions of the left hand side (LHS) are $M^0L^3T^0$ and with units $m^3$.

Dimensions of LHS = $M^0L^3L^0$
Dimensions of RHS = undefined
Units of LHS = $m^3$ (*SI* system)
Units of RHS = undefined (in every system)

Equation (5.27) also defies the definition of the physical quantity. A physical quantity $q$ is expressed as the product of the numerical value of $N$ and a unit of measurement $U$,

$$Q = N \times U$$

or Force ($F$) = 28 dyne

Here 28 is the magnitude of force and dyne is the unit. But the equation of the volume, i.e. Eq. (5.27), possesses neither the numerical value nor unit. So the identity of volume vanished in this case when the Archimedes principle is applied. But the results are consistent if the generalized form of the Archimedes principle is used as in Eq. (5.37)
But Eq. (5.27) which is derived from it is dimensionally inconsistent. The dimensional homogeneity is the first and

foremost condition for consistency of an equation and all equations have to obey it without exception. If any equation does not obey it then that equation loses its identity. This is a serious limitation of the equation based upon the Archimedes principle as it disobeys the dimensional homogeneity and definition of physical quantity.

## *2.4. Division by zero smoothly follows from equations based on the Archimedes principle*

The division by zero is not allowed, and this conclusion serves as a restriction for scientists that such equation should not be formulated when division by zero is not feasible. If scientists formulate such equation then it is invalid; even Einstein did so, and later accepted it as a mistake. Under some conditions, the situation with the Archimedes principle is similar, it also involves division by zero. Every equation needs to be checked for its consistency for various values of the involved parameters. It is the established method for all observations. Although division by zero is not permitted, yet it smoothly follows from the equations based upon Archimedes' principle as per its definition. In this case, the numerator of the equation also becomes zero. So it is the only limitation of the Archimedes principle. Only logical and feasible analysis is carried out. So it is not the fault of interpretation but that of the law which is being interpreted. The logical mathematical analysis cannot be ignored and to inherent limitations of the equation must be highlighted. It is the scientific approach.

This situation can be obtained in number of ways for various media. This equation in this particular case is not applicable [9]. The problem can only be solved by generalizing the 2265 years old Archimedes principle. Even after the enunciation of $g$ in 1685, this equation, i.e. Eq. (5.27) was reported after 327 years. Even then the results were reported by the author for first time. Such specific studies do not mean any comment or conclusion regarding the established status of the principle in other cases.

### *Existing literature*

Archimedes did not check this aspect of the principle. The reason is very simple – these equations did not exist at that time. Now these equations have been feasible for the past 327 years (since 1685 when Newton published the Principia). In the existing literature there is no evidences that Archimedes extended the application of his principle to floating bodies due to the lack of mathematical background. Such application may have been qualitatively extended by other scientists.

In the standard existing literature S. Borowitz and A. Beiser [6] quoted similar equations but neglected the value of $v$.

According to the existing literature as quoted by Borowitz and Beiser, the weight of the balloon in which volume $V$ is filled with the gas is

Weight of air-filled balloon
= mass of balloon × acceleration due to gravity
= [mass of air-filled balloon + mass of sheath] $g$
$$= (VD_a + m)g \tag{5.28}$$

Upthrust experienced by air-filled balloon
= Volume of air-filled balloon × density of water × acceleration due to gravity

$$U = VD_w g \tag{5.8}$$

Let the balloon be filled with air and rises upward in water as the upthrust dominates. When the weight and upthrust are balanced then the balloon floats.

weight of air-filled balloon = upthrust exerted by water

$$(VD_a + m)g = VD_w g \tag{5.29}$$

$$m = VD_w - VD_a \tag{5.30}$$

Up to this stage the equation quoted by Borowitz and Beiser [6] says that greater the volume of helium in the balloon, the higher the mass it supports. For this case Eq. (5.30) can be written as

$$m = VD_a - VD_{he}$$

or
$$V = \frac{m}{(D_a - D_{he})}$$

The helium-filled balloon floats in air when the density of air is equal to that of helium filling the balloon, i.e. $D_a = D_{he}$, in that case the volume of the balloon becomes infinite, i.e.

$$V = \infty$$

This prediction is not justified at all, as it contradicts the law of conservation of matter. The balloon with a volume of 50 cc or less or more can float in air. Borowitz and Beiser neglected value of $v$

in equations, which is not correct. So equations are to be completely checked before drawing conclusions.

The situation as in Eq. (5.27) can be obtained in a number of ways for various media. This equation is not applicable in this particular case [9]. This problem can only be solved by generalizing the 2265 years old Archimedes principle. Even after the enunciation of $g$ in 1685, this equation, i.e. Eq. (5.27), has been reported after 327 years. Such specific studies do not mean any comment or conclusion of the established status of the principle in other cases. But the universality and generality of the principle are definitely restricted.

## 2.5. *Numerical calculations of volume V*

Let the internal volume of a balloon or vessel be 200 cc and filled with platinum (21.5 gm per cc) with a volume of 2cc, silver (10.5 gm per cc) with a volume of 2cc, cork (0.24 gm per cc) with a volume of 10 cc, and the rest is air and metal. The density of the medium is equal to that of water in which the balloon floats. Let the volume of the sheath be 20 cc. The sheath is made of a non-hygroscopic material and floats in water.

The sheath may be non-hygroscopic if its composition is 93.8% in weight Nylon 12 and 6.2% Santicizer [14]. Now in this case $V$ = 200 cc, $v$ = 20 cc, $D_w = D_m$; then the mass of the sheath can be calculated from Eq. (5.14)

$$m = (V + v)\, D_w - VD_m = (200 + 20) - 200 = vD_w = 20 \text{ gm} \tag{5.31}$$

It is evident that when $D_w = D_m$, then the mass of the sheath with a volume of 20 cc must be 20 gm. So the density of the sheath must be equal to 1 gm per cc. It is incorrect to argue that the mass which the balloon supports is independent of volume $V$, i.e. the volume of the medium (Pt, Ag, cork, metal and air). But it can be justified with the help of logical deductions that this interpretation is not correct or logical. The mass of the sheath of 20 gm supports the volume of 200 cc having a density of 1 gm and floats in water of the same density.

(a) Equation (5.31) may not be regarded as true for every volume. It is derived under the conditions; $V$ = 200 cc, $v$ = 20 cc, $m$ = 20 gm, $D_m = D_b$ = 1 gm per cc. Now Eq. (5.31) is independent of volume $V$. It must not be regarded true for all volumes. The volumes may be 0.1 cc or less and

$2 \times 10^6$ cc or $2 \times 10^{20}$ cc or more, the equation cannot hold for all cases. If volume $V$ of the medium (Pt, Ag, cork, metal and air) filled inside the sheath of volume 20 cc is $2 \times 10^{20}$ cc (any volume). The magnitude of the medium, i.e. $2 \times 10^{20}$ cc, can never be contained in a sheath of volume 20 cc and mass 20gm. Obviously the sheath can never be able to support or accommodate the volume of $2 \times 10^{20}$ cc and hence will break down. The constituents Pt, Ag, Al and metal fall down; metal and air will rise upward. Thus, it does not represent the actual experimental situation that the balloon floats completely submerged in the fluid.

(b) Under the given conditions the volume $V$ cannot be regarded as true for all volumes, e.g. $10^5$ cc or more and $10^{-5}$ cc or less. How the constituents of the balloon, i.e. platinum (21.5 gm per cc) of volume 2 cc, silver (10.5 gm per cc) of volume 2 cc, cork (0.24 gm per cc) of volume 10 cc and the rest is air and metal will be condensed to 0.1 cc? Hence the deduction that the volume equal to 200 cc (filled in the balloon) will reduce to 0.1 cc is not justified. Earlier, this volume was supposed to be condensed in a volume of 20 cc. If the mass equal to 200 gm is confined to a volume of 0.1 cc, then it will have the density of 2000 gm per cc. So the initial deduction that Eq. (5.31) is true for every volume is incorrect. Thus, the equation hold good only for those values of volume $V$, for which it is defined. The volume cannot be assigned arbitrarily.

Therefore. the numerous values cannot be assigned to $V$ which is 200 cc in this particular system. Hence the original form of Archimedes' principle is not applicable in this particular case. Under this condition the volume $V$ can be calculated from Eq. (5.18) as

$$V = \frac{(m - vD_w)}{(D_w - D_m)} \tag{5.18}$$

Putting the value of mass from Eq. (5.31), $V = 200$ cc, $v = 20$ cc, $D_m = D_w = 1$ gm per cc in Eq. (5.18), we get

$$V = \frac{(20-20)}{(1-1)} = \frac{0}{0}$$

which is undefined and thus does not signify anything. This is the one and only limitation of the Archimedes principle. In his treatise

**Table I.** The values of $v$, $D_m$, $D_w$ and $V$ of the floating balloon under condition $D_m = D_w$ based on the original form of the Archimedes principle, i.e. $U = D_m V_g$

| Sr. No | Equations | Result |
|---|---|---|
| 1 | $v = \left\{\dfrac{m + V(D_m - D_w)}{D_w}\right\}$ | $v = v$ (correct) |
| 2 | $D_w = \dfrac{(m + VD_m)}{(V + v)}$ | $D_w = D_m$ (correct) |
| 3 | $D_m = \left\{\dfrac{D_w(V + v) - m}{V}\right\}$ | $D_w = D_m$ (correct) |
| 4 | $V = \dfrac{(m - vD_w)}{(D_w - D_m)}$ | $V = \dfrac{0}{0}$ (undefined) |

Note: The generalized form of the Archimedes principle predicts that volume becomes indeterminate, which is not justified.

Introduction to Fluid Dynamics, G.K. Batchelor [10] discussed floating bodies under a separate heading. However, he did not discuss this case at all. So this problem has not been studied in the existing literature.

The various values of the variables based upon the original form of the Archimedes principle are shown in Table I. The various values of variables based upon the generalized form of the Archimedes principle are shown Table II.

## 3.0. Generalization of the Archimedes principle

Archimedes stated the principle for different purposes, from observations which had been witnessed by numerous people. But its applications were then extended to floating bodies. It is confirmed in Eq. (5.27) that under some conditions, the volume becomes indeterminate (V = 0/0), which is not justified. However, this deduction does not apply to other cases where the Archimedes principle is quantitatively confirmed. Practically in this case the weight of the balloon, artefact or vessel decreases in the fluid. Thus, the principle has to be validated for all conditions; it can be so if its definition is generalized. The principle cannot be applied only to selective cases: 'when body is wholly or partially immersed in the fluid at rest then its weight decreases (experiences buoyant force) and the decrease in

**Table II.** The values of $v$, $D_m$, $D_w$ and $V$ of the floating balloon under condition $D_m = D_w$ based on the generalized form of the Archimedes principle. i.e. $U = f D_m V g$, where $f$ is the coefficient of proportionality

| Sr. No | Equations | Result |
|---|---|---|
| 1 | $v = \left\{ \dfrac{m + V(D_m - D_w)}{D_w} \right\}$ | $v = v$ (correct) |
| 2 | $D_w = \dfrac{(m + VD_m)}{f(V + v)}$<br>$D_w = \dfrac{(m + VD_m)}{(V + v)}$<br>$D_w = \dfrac{(m + VD_m)}{(V + v)}$ | $D_w = D_m$ (correct) |
| 3 | $D_m = \left\{ \dfrac{fD_w(V + v) - m}{V} \right\}$ | $D_w = D_m$ (correct) |
| 4 | $V = \dfrac{(m - fvD_w)}{(fD_w - D_m)}$ | $V = V$ |

Note: The generalized form of Archimedes principle predicts that $V = V$, which is justified.

the weight of the body in the fluid is proportional to the weight of the fluid displaced.'

The experimental observation that the weight of the body decreases in the fluid is correct, but the magnitude of the decrease in weight can be slightly less or more than predicted by the principle. It may be due to unaccounted factors such as the shape of the body, the viscosity of media. Currently it is inference on the basis of specific observation mentioned above. In this particular case the equation can then give logical results. This observation, as described above, is different from weighing the body in fluids, as in those cases the body was attached to the balance. In this case the weight of the body is independently balanced by the upthrust. The generalized form of the principle [2] is 'the upthrust experienced by the body in the fluid is proportional to the weight of the fluid displaced'

$$U_{gen} \propto (V + v)\, D_w\, g \text{ or } U_{gen} = f\,(V + v)\, D_w\, g \qquad (5.32)$$

where $f$ is the coefficient of proportionality. The magnitude of $f$ is determined experimentally like all other coefficients of proportionality in existing physics, e.g. the coefficient of viscosity, the coefficient of thermal conductivity.

(i) If $f = 1$, then both original and generalized forms of the principle are the same.

$$\text{Original form } U = (V + v)\, D_w\, g \tag{5.13}$$

$$\text{Generalized form } U_{gen} = f\,(V + v)\, D_w\, g \tag{5.32}$$

(ii) If $f > 1$

$$U_{gen} > U$$

or $$f\,(V + v)\, D_w > (V + v)\, D_w g$$

(iii) If $f < 1$

$$U_{gen} < U$$

or $$f\,(V + v)\, D_w < (V + v) D_w g$$

Now the upthrust can be slightly or infinitesimally less or more than the weight of the fluid displaced, depending upon the value of $f$. The value of $f$ can be determined experimentally. The coefficient of proportionality $f$ takes those factors or parameters into account which are not accounted for by the Archimedes principle, e.g.

(a) shape of floating body,
(b) viscosity of medium
(c) magnitude of medium
(d) surface tension, etc.

Now the volume of the medium filling the balloon becomes exact, i.e. $V$.

**The generalized form predicts exact volume $V = V$**

Now all the calculations can be done for all those cases, as in the case of the original form of the Archimedes principle. The consistent results can be obtained in this regard.

The body floats in the medium if the weight of the body is equal to the upthrust, i.e.

Weight of body = Generalized upthrust

$$V D_m\, g + mg = f\,(V + v)\, D_w\, g \tag{5.33}$$

or $$m = f\,(V + v)\, D_w - V D_m \tag{5.34}$$

$$V = \frac{(m - fvD_w)}{(fD_w - D_m)} \tag{5.35}$$

If $f = 1$ then Eq. (5.35) is the same as Eq. (5.18).

The various other values of parameters, i.e. $v$, $D_m$, $D_b$, can be written as

$$v = \left\{\frac{m + V(D_m - fD_w)}{D_w}\right\}$$

$$D_w = \frac{(m + VD_m)}{f(V + v)}$$

$$D_m = \left\{\frac{fD_w(V + v) - m}{V}\right\}$$

The density of water is assumed equal to the density of the medium, i.e. ($D_m = D_w$), then Eq. (5.34) becomes

$$m = fVD_w + fvD_w - VD_w \tag{5.36}$$

In the case of the original form of the Archimedes principle under this condition the value of $m$ was

$$m = vD_w \tag{5.20}$$

Thus, in the case of the generalized form of the Archimedes principle, the value of $m$ varies.

Now substituting Eq. (5.34) in Eq. (5.35) and $D_m = D_w$ we get

$$V = \frac{(f(V + v)D_w - VD_w - fvD_w)}{(fD_w - D_m)} = \frac{(f-1)D_w V}{(f-1)D_w} = V \tag{5.37}$$

Now consistent results are obtained for the value of $f$ different from unity. Also correct values of $v$, $D_m$ and $D_w$ are obtained. Unlike Eq. (5.18), in this case, division by zero is not involved and the numerator is non-zero, hence consistent and logical results are obtained. The dimensions and units are the same in both LHS and RHS, i.e. $M^0L^3L^0$ and $m^3$. This equation is dimensionally consistent:

Dimensions of LHS = $M^0L^3L^0$
Dimensions of RHS = $M^0L^3L^0$
Units of LHS = $m^3$ (*SI* system)
Units of RHS = $m^3$ (*SI* system)

Therefore, both sides are dimensionally consistent and have the same units. Thus, the generalized equation is mathematically consistent. Then these equations, based upon the generalized form of the Archimedes principle, are mathematically consistent and do not have any limitation like the equations based upon the Archimedes principle. Hence mathematically the generalization is justified. The value of the coefficient of proportionality $f$ can be determined experimentally.

Correctly, from the generalized form of the principle, the internal volume of the balloon is $V$ which contains platinum, silver, cork, metal and air. In this particular case the theoretical derivation of the Archimedes principle is consistent, if it is generalized, thus the coefficient of proportionality $f$ comes in picture. The magnitude of the coefficient of proportionality can be checked experimentally. It can be understood that $f$ has physical significance or experimental validity and takes those factors into account which are not accounted for by the Archimedes principle. These factors can be considered as

(i) Shape of body
(ii) Viscosity of fluid
(iii) Magnitude of medium
(iv) Surface tension of medium etc.

According to the Archimedes principle the body floats in the fluid if the buoyant force is equal to the weight of the fluid displaced [11, 12]. Further, the density of the body ($D_b$) is equal to the density of the medium($D_m$), as in Eq. (5.11). Apart from densities of the body and the medium, the shape of the body, the viscosity of the fluid, the size of the medium, surface tension, etc. are the factors which can influence the results. Thus, the coefficient of proportionality $f$ in the generalized form of the Archimedes principle depends upon these factors i.e. the shape of the body, the viscosity of the fluid, ?the size of the medium, surface tension, etc. Practically, the value of the coefficient of proportionality $f$ can be determined experimentally. As the Archimedes principle also determines the conditions of sinking and rising [13], so the coefficient of proportionality can influence such results also. It can be specifically studied.

## 4.0. Mathematical derivation of the Archimedes principle is only true for symmetric shaped bodies. It is not true for body of any shape

The mathematical form of the Archimedes principle is

$$\text{Upthrust} = \text{buoyant force} = \text{weight of fluid displaced}$$
$$= VDg \tag{5.8}$$

where $V$ is the volume of the body, $D$ is the density of the liquid and $g$ is the acceleration due to gravity.

It can be easily conducted from its theoretical derivation that this derivation is only true for the body of a particular shape. The theoretical derivation is not true for any shape.

Consider a body of mass $M$ completely immersed in a liquid of density $D$ and being under static equilibrium with the liquid [5, 6]. Let $A$ be the area of the cross section of the upper and lower surfaces of the body. Thus the area of both surfaces is precisely equal. Let $h$ be the height of the body. As this body is immersed in the liquid, let $L$ be the depth of the upper face of the body from the free surface of the liquid.

The pressure due to the height of the liquid column $H$ is given by, $P = DgH$. The downward pressure exerted on the upper surface of the body is

$$P_1 = LDg \tag{5.38}$$

The upward pressure acting on the lower surface of the body

$$P_2 = (L + h)\ Dg \tag{5.39}$$

The downward force on the upper surface of the body is

$$F_1 = P_1A = LDgA \tag{5.40}$$

Similarly, the upward force on the lower surface of the body is

$$F_2 = P_2A = (L + h)\ DgA \tag{5.41}$$

Therefore, the net upward thrust on the body is given by

$$F = F_2 - F_1 = (L + h)\ DgA - LDgA = AhDg \tag{5.42}$$

But, $Ah$ = Area × height = $V$ = volume of cylinder

$$\text{Upward thrust} = F = U = VDg \tag{5.8}$$

## *4.1. Condition of derivation of upward thrust*

The equation for the upthrust is derived if the areas of the upper and lower surfaces are precisely equal.

Area of upper surface = Area of lower surface

If the areas are not equal then the result is not

$$\text{Upward thrust} = F = U = VDg \tag{5.8}$$

Thus the derivation of the upthrust is only true for the body of a definite shape. This derivation is not true for bodies of every shape. However, the Archimedes principle was stated for measurement of the density of the body of an irregular shape. It is regarded as true for every shape. Definitely, it is due to the reason that when in 250 BC Archimedes stated the principle there was no basis for mathematical equations. This derivation became feasible 1935 years after enunciation of the Archimedes principle when Newton published the law of gravitation in 1685.

## 5.0. Physical significance of the coefficient of proportionality

The applications of the Archimedes principle are extended to rising, falling and floating bodies. However the Archimedes principle was initially established to determine density by attaching a body to a balance with the help of a string for weighing. But the rising and falling bodies are not externally attached to the balance or instrument. In the case of independently floating bodies only two forces are considered, i.e. weight and upthrust. Both these forces balance each other. This is the conceptual difference between original the verification of the Archimedes principle and its applications. The idea of the coefficient of proportionality was originated from mathematical analysis of the principle in the case of completely submerged floating balloons. $f$ takes into account the factors which are not taken into account by the original form of the principle but can potentially influence the results. The perception of $f$ can be justified or understood

**Table III.** The various characteristics of water, glycerine and mercury

| Characteristic | Water | Glycerine | Mercury |
|---|---|---|---|
| Density | 1 | 1.26 | 13.6 gm/cc |
| Coeff. Of viscosity (poise) | $1.01 \times 10^{-2}$ | 10.69 | $15.5 \times 10^{-3}$ |
| Surface tension | 75.6 | 63.1 | 465 (dyne/cm) |
| Angle of contact | 8–9 | — | 137 |
| Capillarity | Rise | — | Fall |
| Physical behaviour | Wets | Wets | Does not Wet |
| Height of liquid column (h) | 10.33 | 8.202m | 0.76m |

in a number of ways as the principle is also applied in explaining the rising and falling bodies. The various results about the generalized form of the Archimedes principle are shown in Table III.

## *5.1. Floating bodies*

According to Archimedes, for floating bodies only the densities of bodies and media are relevant, and the rest of factors, e.g. shape of the body, magnitude, surface tension, viscosity of medium, etc. are irrelevant. In terms of the generalized form of the Archimedes principle, the weight of the body is equal to the upthrust exerted by the medium (buoyant force)

$$VD_b g = f\,VD_m\,g$$

or

$$D_b = f\,D_m \tag{5.43}$$

Thus when the body floats then $f$ takes into account other factors (other than $D_m$ and $D_b$), e.g. shape of body, magnitude, surface tension, viscosity of medium etc. Thus, theoretically, the generalized form of the Archimedes principle is complete, as it takes all possible involved factors in account and the original form takes in account $D_b$ and $D_m$ only. The effect of these factors in such phenomena can be confirmed in specifically designed experiments. It can be confirmed by fabricating bodies of non-hygroscopic nature [14] of different shapes (flat or umbrella-shaped or distorted shape), if such typical bodies float in water of slightly higher density then the value of $f$ other than unity will be confirmed. Now experiments in the most viscous medium – glycerine – and dense medium –mercury – are considered along with water. The various characteristics of liquids are shown in Table III.

**(i) Experiments in glycerine:**

The glycerine may be regarded the most viscous medium in such experiments. Let us fabricate a vessel (balloon) which has the internal volume of 200 cc and the volume of the sheath is equal to 20 cc. Let the vessel be internally filled with suitable amounts of platinum, silver, cork, metal and air such that the overall density inside the medium ($D_m$) is equal to 1.26 gm per cc. Thus, the internal volume of the balloon = 200cc

Density of medium inside balloon = 1.26 gm per cc

Mass of medium inside the balloon = 252gm

Now, according to the original form of the Archimedes principle, i.e. from Eq. (5.14), the mass of the sheath should be 25.2 gm, thus the density of the sheath is also equal to 1.26 gm per cc. Thus, the density of the balloon is precisely equal to that of the medium:

$$m = (V + v)\, D_{hg} - VD_m \tag{5.14}$$
$$= (200 + 20)\ 1.26 - 200 \times 1.26$$
$$= 77.2 - 262 = 25.2 \text{ gm} \tag{5.44}$$

Therefore, according to the original form of the Archimedes principle the balloon floats in glycerine if its density is equal to that of glycerine, i.e. 1.26 gm per cc. The density of the balloon is precisely equal to that of the medium.

**Possible deviation from the Archimedes principle may result from the following observation:**

Now, let a balloon ($V$ =2.00 cc; $v$ =20 cc, $m$ = 25.21 gm) of arbitrary shape (typically flat or umbrella shaped or typical body) floats in glycerine (may be in cup, jug or tank such that its density is 1.26 gm per cc) at any depth. It implies that the mass of the sheath of the balloon is slightly more, i.e. 25.21 gm, than the mass predicted by the Archimedes principle, i.e. 25.2 gm. If actually under this situation the balloon floats then there would be a deviation from the Archimedes principle and the value of $f$ can be calculated from Eq. (5.14)

$$m = f\,(V + v)\ D_{gly}\ VD_{gly} \tag{5.34}$$

$$25.21 = f\,(220)(1.26) - (200) \times 1.26 \text{ or } f = 1.00003607 \tag{5.45}$$

Thus, the deviation from the original form of the Archimedes principle can be observed. The value of $f$ will be 1.000036075 from Eq. (5.45). Thus %age deviation from the Archimedes principle will be $3.6 \times 10^{-3}$. It is simply the theoretical description of the experiment and the final results will depend upon sensitive experiments and repeatability. These experiments are also significant in the sense that they are not specifically and quantitatively conducted by the scientists. Whatever may the result be Eq. (5.34) is capable of explaining it. This experiment can be repeated for other bodies (of different shapes) having the density slightly different than that of the medium.

(ii) **Experiments in mercury:** Mercury is the densest fluid and the bodies can also float in mercury. Thus, for completeness the sensitive experiments can be conducted in mercury. Let us fabricate a vessel which has an internal volume of 200 cc and the volume of the sheath is equal to 20 cc. Let the vessel be filled with a suitable amount of the medium, as in the previous case, so the overall density of the medium inside is equal to 13.6 gm per cc.

Internal volume of balloon = 200 cc
Density of medium inside balloon = 13.6 gm per cc
Mass of medium inside balloon = 272 gm

Now, according to the original form of Archimedes' principle, i.e. from Eq. (5.14) (equivalently written for this situation) the mass of the sheath should be 272 gm per cc (such that the density of the sheath should be 13.6 gm per cc)

$$m = (V + v)\, D_{hg} - VD_{hg} \tag{5.14}$$

$$= (200 + 20) \times 13.6 - 200 \times 13.6$$

$$= 2992 - 2720 = 272 \text{ gm} \tag{5.46}$$

Thus the density of the balloon is precisely equal to that of the medium. The balloon or vessel must float the supporting mass equal to 272 gm. Therefore, the density of the balloon is precisely equal to that of the medium. Any deviation from this observation would favour the generalized form of the Archimedes principle.

**Possible deviation from the Archimedes principle may result from the following observation:**

In this case it is assumed the balloon floats the supporting mass equal to slightly more than 272 gm, i.e 272.5 gm. Now let a vessel ($V$ = 200 cc, $v$ = 20 cc and $m$ = 272.5 gm) of any arbitrary shape float in mercury (may be in a cup, jug or tank) with the density of 13.6 gm per cc experimentally. Under these conditions, Eq. (5.34) becomes

$$272.5 = f\,(200 + 20) \times 13.6 - 200 \times 13.6$$

$$= f\,2992 - 2700$$

or $$f = \frac{2992.5}{2992} = 1.00016711 \tag{5.47}$$

Then the value of $f$ from Eq. (5.47) will be 1.000167 as in the case of Eq. (5.45). Thus, the %age deviation from the original form of the Archimedes principle is $1.67 \times 10^{-2}$. However, all observations are consistent with the generalized form of the Archimedes principle. This experiment can be repeated for other bodies (of different shapes) having density slightly different than that of the medium.

**(iii) Experiments in water:**

Let us fabricate a vessel (balloon) which has the internal volume 200 cc and the volume of the sheath is equal to 20 cc. Let the vessel be internally filled with suitable amounts of platinum, silver, cork, metal and air such that the overall density of the inside medium ($D_m$) is equal to 1 gm per cc. Now, according to the original form of the Archimedes principle, i.e. from Eq. (5.14) the mass of the sheath should be 20 gm, thus the density of the sheath is also equal to 1 gm per cc. Thus, the density of the balloon is precisely equal to that of the medium.

$$m = (V + v)\, D_w - VD_m \tag{5.14}$$
$$= (200 + 20) - 200$$
$$= 220 - 200 = 20 \text{ gm} \tag{5.48}$$

Thus, according to the original form of the Archimedes principle the balloon floats in water if its density is equal to that of water, i.e. 1 gm per cc. The density of the balloon is precisely equal to that of the medium.

**Possible deviation from Archimedes principle may result from the following observation:**

Now, let a balloon ($V = 2.00$ cc; $v = 20$ cc, $m = 1$ gm) of an arbitrary shape (typically flat or umbrella-shaped) floats in water (may be in a cup, jug or tank such that its density is 1 gm per cc) at any depth. It implies that the mass of the sheath of the balloon is slightly more, i.e. 25.21 gm, than the mass predicted by the Archimedes principle, i.e. 20.001 gm. If actually under this situation the balloon floats then it would be a deviation from the Archimedes principle, and the value of $f$ can be calculated from Eq. (5.34)

$$m = f\,(V + v)\,D_w - VD_w \tag{5.34}$$

$$20.001 = f\,(220)(1.26) - (200)1.26$$

$$\text{or } f = 1.000004545 \tag{5.49}$$

Thus, a deviation from the original form of Archimedes' principle can be observed. The value of $f$ will be 1.000004545 from Eq. (5.49). Thus, the % age deviation from Archimedes' principle will be $4.545 \times 10^{-4}$. It is simply a theoretical description of the experiment and the final results will depend upon sensitive experiments and their repeatability. These experiments are also significant in the sense that they have not been specifically and quantitatively conducted by the scientists yet. Whatever may the result be Eq. (5.14) is capable of explaining it. This experiment can be repeated for other bodies (of different shapes) having a density slightly different than that of medium.

## 5.2. *Independent experimental support to above experiments*

***The falling bodies***

All the bodies fall with constant acceleration in vacuum. In fluids the acceleration of the body decreases. Gorge Gabriel Stokes in 1845 put forth that bodies fall with constant velocity in fluids under some conditions or postulates [6, 15], e.g. the body must be small and spherical in shape, the medium should be large in extent, the fluid must be viscous in nature, etc. Thus, the shape of the body is considered here and is spherical. Mathematically Stokes calculated the equation for constant velocity as

$$c = \frac{2r^2\left(1 - \dfrac{D_m}{D_b}\right) g D_b}{9h} \tag{5.50}$$

where $c$ is constant velocity, $r$ is the radius of the small sphere, $D_m$ is the density of the medium, $D_b$ is the density of the body and $\eta$ is the coefficient of viscosity.

Equation (5.50) is obtained with te help of

(i) the Archimedes principle, $U = VD_m\, g$

(ii) law of gravitation $W = mg$

(iii) viscous force $F = 6\pi\eta rc$

under certain assumptions.

Equation (5.50) is discussed extensively in Chapter 6. The Stokes law finds numerous applications in fluid dynamics. After 65 years, in 1910 Arnold [15] verified Eq. (5.50) in water with an accuracy of a few tenths of 1% for a sphere of rose metal with a radius of 0.002 cm, i.e. $V = 33.524 \times 10^{-9}$ cc. Thus, it is applicable in an extremely narrow range. Many scientists have generalized or extended it. The small spheres attain constant velocity due to the viscous force of the fluid. Thus, the shape of the body is also a factor which influences the results in motion of bodies; these observations are consistent with the generalized form of the principle. Such experiments can also be conducted with rising bodies.

## 5.3. How the shape of the body is important in falling bodies

The equation of constant or terminal velocity, i.e. (5.50), is based upon the assumption laid down by Stokes that the body must be spherical. This equation has been experimentally confirmed in a limited range. So the shape of the body has a significant role to play in falling bodies. Equation (5.50) has not been confirmed for flat or distorted shaped bodies. We have the generalized form of the Archimedes principle as

$$U = f\,(V + v)D_w\, g \tag{5.34}$$

where $f$ is the constant of proportionality. It takes the shape of the body into account together with other factors. Thus, the generalized form of the Archimedes principle, which takes the effect of the shape into account, is justified in this case. For example, only spherical bodies (particular shape) are found to fall with constant velocity not taken into account by the Archimedes principle. So the shape is a significant factor which is not taken into account by the Archimedes principle.

## 5.4. Accelerated motion

The Stokes law, which confirms the effect of the shape of the body and the viscosity of the medium, is applicable over a narrow range and bodies move with a constant velocity. Thus, the spheres with a radius of more than 0.002 cm fall with a variable velocity or motion is accelerated. Such experiments are not reported in the literature. If such experiments are conducted, then range of validity of the Stokes law can be assessed over a wide range. These experiments will establish the effect of the shape in downward motion of bodies and the extent of validity of the Stokes law over a wide range. Indirectly in such experiments the generalized form of the Archimedes principle will also be tested. The weight, upthrust and hence the resultant weight depend only upon the volume $V$, $D_m$ and $D_b$. The factors like the shape of the body and the viscosity of the medium are elusive to the Archimedes principle as clear from Eq. (5.8), i.e. $U = VD_m g$. It is the physical significance of the generalized form the of Archimedes Principle.

## 5.5. *Downward motion of bodies when the Stokes law is not applicable*

The Archimedes principle implies that a body sinks in a medium if the resultant weight is positive, i.e. the density of the body is more than that of the medium ($D_b > D_m$). This aspect is discussed in Chapter 7. The resultant weight is given by

$$W_R = (D_b - D_m)\, Vg = (1 - \frac{D_m}{D_b})D_b Vg \qquad (5.52)$$

Further, the resultant acceleration can be written as

$$a_R = (1 - \frac{D_m}{D_b})g \qquad (5.53)$$

In general, if a flat steel sheet (5 m × 5 m or distorted shape) with a mass of 20 gm or a spherical steel body with a mass of 20 kg is dropped in water, the spherical body falls more quickly than the flat sheet of the 20 gm mass. Both the steel spheres and the flat sheet have the same resultant acceleration and should fall equal distances in equal time, according to the Archimedes principle. The reason is that both resultant accelerations are the same for both bodies. The resultant weight for the two bodies is

$$W_R = (D_b - D_m)Vg = (1 - \frac{D_m}{D_b})D_b Vg \qquad (5.54)$$

$$a_R = (1 - \frac{D_m}{D_b})g = (1 - 1/7.8) \times 980 = 854.35 \text{ cm/s}^2$$

$$S = \frac{1}{2}a_R t^2 = \frac{1}{2}(1 - \frac{D_m}{D_b})g\, t^2 \qquad (5.55)$$

The resultant acceleration and distance travelled depends upon density of body only. Hence the bodies (flat of mass 20 gm and

**Table IV.** Consequences of the indeterminate form of volume

| Sr. No. | Characteristic | Original form of Archimedes principle | Generalized form of Archimedes principle |
|---|---|---|---|
| 1 | Definition | $U = VD_m g$ | $U = fVD_m g$ |
| 2 | Volume under certain conditions | $V = \frac{0}{0}$ | $V = V$ (500 cc, say) |
| 3 | Coefficient of proportionality | $f = 1$ | $f < 1$ or $f > 1$ or $f = 1$ |
| 4 | Status of parameters | Does not account for shape of body, viscosity of medium, etc. | Takes into account the shape of body, viscosity of medium and other factors. |
| 5 | Specific experiments | Breaks down under certain conditions | Some specific experiments are suggested to determine effects of shape and viscosity of medium for completely submerged floating bodies. Typical experimental value of $f$ is assessed as1.000008. |

spherical of mass 20 kg and density 7.8 kg/m$^3$) should fall equal distances in equal intervals of time.

But the flat body or sheet falls more slowly than the spherical body. There is no factor in Eq. (5.10) which may account for the shape of the body. It is due to the shape of the body. In this case the effect of viscous force is negligible and is only applicable in a limited range. Similar results can be obtained if bodies of different masses in different media are studied. Can a similar explanation be identically given for rising bodies? It will add to scientific knowledge in either way. Thus, various observations can be explained with the help of the generalized form of the Archimedes principle. The applications of the Stokes law can be extended to rising bodies as well but it is not the theme of the current discussion. The various results of the original and generalized form the of the Archimedes principle are shown in Table IV.

## Frequently asked questions

1. **Why did other scientists not point out that under some conditions the Archimedes principle predicts that mass becomes indeterminate ?**

   **Answer:** The deduction that the volume becomes indeterminate is based upon mathematical equations.

$$V = \frac{\left((V+v)D_w - D_w V - vD_w\right)}{\left(D_w - D_w\right)} = \frac{\left(vD_w - vD_w\right)}{\left(D_w - D_w\right)} = \frac{0}{0} \tag{5.27}$$

   (i) Archimedes stated the principle in about 250 BC, i.e. about 2265 years ago.

   (ii) At that time there were no equations based upon the Archimedes principle as we do have now.

   (iii) Sir Isaac Newton published the Principia in 1685, propounded the Universal Law of Gravitation and defined acceleration due to gravity *g*. Thus, weight and upthrust were defined. It happened 1935 years after the enunciation of the Archimedes principle. So it was beyond the remit of Archimedes to critically analyze his principle in this regard in the absence of mathematical background.

   (iv) Weight, $W = mg$ and upthrust, $U = VDg$ were then defined. Even now these equations are part of scientific literature since in the past 327 years none has pointed out this limitation of the principle, i.e. Eq. (5.2).

   (v) The famous physics authors S. Borowitz and A. Beiser [6] have quoted very similar equations but neglected the term $v$. Also G.K. Batchelor, in his book An Introduction to Fluid Dynamics, discussed the floating bodies but did not address the issue, i.e. completely submerged flotation of balloons [10].

   Thus, it may be stated that initially due to lack of mathematical equations scientists did not critically analyze the principle in this way. At a later stage when mathematical equations were formulated, scientists thought it is an established topic and no attention was given to it. But scientific queries and advancements always continue.

2. **What are the consequences of the fact that the Archimedes principle predicts that volume becomes indeterminate?**

**Answer:** This mathematical equation indicates limitations of the Archimedes principle. Consequently, the following new outcomes become clear:

(i) the 2365 years old Archimedes principle is generalized as 'when a body is wholly or partially immersed in a fluid at rest then its weight decreases (experiences buoyant force) and the decrease in the weight of the body in the fluid is proportional to the weight of the fluid displaced,'

or

'when a body is wholly or partially immersed in a fluid at rest then its weight decreases (experiences buoyant force) and the decrease in the weight of the body may or may not be equal to the weight of the fluid displaced.'

(ii) It is suggested that the original form of the Archimedes principle does not take the following factors into account.
   (a) Shape of body
   (b) Viscosity of fluid
   (c) Magnitude of medium
   (d) Surface tension of medium, etc.

   Moreover, at the time of Archimedes the concepts of viscosity of fluids and surface tension were not understood or it can be stated that they did not exist.

(iii) For complete understanding of the principle some new experiments are suggested. Thus, this interpretation adds to knowledge, which is important.

(iv) The result $V = 0/0$ indicates limitations of the Archimedes principle and leads to its generalization. The generalized form of the Archimedes principle, $U = fD_m Vg$ predicts the exact volume, i.e.

$$V = V$$

3. **What is the main limitation of the equation when volume becomes undefined?**

**Answer:** A physical quantity is a property of an object that can be measured with a measuring instrument. For example, length, width, time, weight, etc. A physical quantity $Q$ is expressed as a product of the numerical value of $N$ and a unit of measurement,

$$Q = N \times U = \text{numerical value} \times \text{unit}$$

Let us consider the physical quantity as volume

$$\text{Volume} = 12 \text{ m}^3$$

Here, numerical value = 12, init = $m^3$. In one particular case the Archimedes principle predicts

$$V = \frac{0}{0}$$

$$\text{Numerical value} = \text{Undefined}$$
$$\text{Units} = \text{undefined}$$

Thus, the identity of the volume vanishes and it is no longer physical quantity. Hence, this interpretation is not correct from the viewpoint of the original form of the Archimedes principle, whereas from the generalized form of the Archimedes principle we get

$$V = V$$

which has consistent dimensions and units.

## References


1. Heath, T.L., *The Works of Archimedes Principle*, C. J. Clay and Sons, (1897).
2. Sharma, A., Speculations in Science and Technology, **20**, 1997, 297–300.
3. Sharma, A., Acta Ciencia Indica, Vol. XXV P, No. 1, 019–022, 1998.
4. Resnick, R. and Halliday, D., Physics, Part I (Wiley Eastern Limited, New Delhi), 1992, pp. 602–604.
5. Green, C.R., *Technical Physics.* Englewood Cliffs: Prentice Hall, 1984, pp.217–31.
6. Borowitz, S. and Beiser, A., *Essential of Physics.* California: Addison Wesley Publishing Company, 1967, pp. 203–210.
7. Newton, I., *Mathematical Principles of Natural Philosophy,* pp. 19–20, London, 1727, translated by Andrew Motte from the Latin.
8. Abbott, A.F., *Ordinary Level of Physics* (Arnold-Heimann (India), New Delhi, Fourth Edition) 1984, pp. 9–10.
9. Sharma, A., Acta Ciencia Indica, Vol. XXIII, 61–64 (1997).
10. Batchelor, G.K., *An Introduction to Fluid Dynamics*, Cambridge University Press, 2000, pp. 17–22.
11. Rorres, R., *Completing Book II of Archimedes's on Floating Bodies,* The Mathematical Intelligence (Springer Publication House), Vol. 26, No. 3, 2004, pp. 32–42.

6

# Is Stokes Law Applicable for Rising Bodies?

**First glimpse**

- ✓ All bodies fall in vacuum with maximum acceleration $g$ (9.8 m/s$^2$). According to the Archimedes principle bodies fall with reduced acceleration in fluids.
- ✓ The Stokes law (1845) predicts that under some conditions a body falls in fluids with subtle constant velocity. This mathematical background is used in measurement of the viscosity of fluids by the method of falling bodies.
- ✓ Viscous force is given by $F = k\eta rv = 6\pi\ \eta rv = 18.8571\ \eta rv$.
- ✓ Sir George Gabriel Stokes (1819–1903), Lucasian Professor of Mathematics, did not experimentally measure the value of $k$ equal to $6\pi$. He might have given it theoretically.
- ✓ The Stokes law is not confirmed for all materials and liquids and apparently the value of $k$ as $6\pi = 18.8571$ is determined theoretically. Thus, its study is incomplete.
- ✓ The Stokes law is regarded as true for rising bodies when $v$ is negative but the conditions (like postulates in falling bodies) of applicability of the law have not been ascertained theoretically or experimentally.
- ✓ If experimental conditions of applicability are ascertained, then the viscosity of fluids can be measured by the method of rising bodies.
- ✓ Without experiments regarding falling and rising bodies over a wide range of parameters the basic law of fluid dynamics is incomplete.

## 1.0. Viscosity

The word viscosity is derived from the Latin viscum. In everyday terms, viscosity is 'thickness' or 'internal friction'. Thus, water is 'thin', having a lower viscosity, while honey is 'thick', having a higher viscosity. If the viscosity of a fluid is less, then greater is its ease of movement (fluidity). Gases also have viscosity, although it is little harder to notice it under ordinary circumstances. The study of flowing matter is known as rheology (the study of the flow of matter, primarily in the liquid state), which includes viscosity and related concepts. The viscosity of water is $1.01 \times 10^{-2}$ poise and that of glycerine 10.69 poise, thus the viscosity of glycerine is about 1058 times than that of water. The reciprocal of viscosity is fluidity.

Viscosity is a measure of a fluid's resistance to flow. It describes the internal friction of a moving fluid. A fluid with large viscosity resists motion because its molecular makeup gives it lot of internal friction. A fluid with low viscosity flows easily because its molecular makeup results in very little friction when it is in motion.

When water contained in a beaker is set into rotation with the help of a finger or a rod, the water starts rotating in the form of coaxial layers. When we stop stirring, speeds of different layers are seen to decrease gradually and finally water comes to rest. It shows that there must be internal friction between layers of water, which tries to destroy the relative motion between layers. This internal friction is known as viscous drag or backward dragging force. 'The property of liquids or gases by virtue of which an internal resistance or friction comes into play, when the liquid is in motion is called viscosity.'

### *1.1. Newton's equation of viscosity*

Sir Isaac Newton (1642–1727) maintained that the viscous force depends upon the velocity gradient, i.e. the rate of change of velocity with distance. Consider a fluid flowing over a horizontal plane of a solid surface in the form of parallel layers. The layer in contact with the solid surface is found to be at rest and as we move up the velocity of layers goes on increasing. Further the uppermost layer possesses the maximum velocity.

Let two layers at distance $x$ and $x + dx$ from the solid surface move with velocity $v$ and $v + dv$. Let the area of each layer be $A$. Then $dv/dx$ denotes the rate of change of velocity with distance and is known as the velocity gradient. Due to viscosity, a force $F$ acts in the opposite direction to destroy relative motion.

According to Newton, the viscous drag force $F$ depends upon the following factors.

1. It is directly proportional to the area of layers in contact

$$F \propto A$$

2. It is directly proportional to the velocity gradient, $\frac{dv}{dx}$

Combining above two proportionalities, we get

$$F \propto A \frac{dv}{dx}$$

$$F = -\eta A \frac{dv}{dx} \tag{6.1}$$

where $\eta$ is the coefficient of viscosity. It depends upon the characteristics of the body and the liquid along with other factors. The negative sign indicated that the viscous force is directed opposite to that of the direction of motion.

$$\text{If } A = 1 \text{ and } \frac{dv}{dx} = 1$$

Then from Eq. (6.1) we get

$$F = \eta \tag{6.2}$$

or viscous drag force = coefficient of viscosity

Hence, the coefficient of viscosity of a liquid may be defined as the tangential viscous force which maintains eq. unit velocity gradient between two parallel layers, each of unit area.

Thus the coefficient of viscosity is a measure of backward dragging force or viscous force.

In the CGS system, the unit of $\eta$ is dyne s $cm^{-2}$ or g $cm^{-2}s^{-1}$, or is called poise.

In the SI system, the unit of $\eta$ is N s $m^{-2}$ or kg $m^{-2}$ $s^{-1}$ or is called decapoise

1 decapoise = N s $m^{-2}$ = $10^5$ (dyne) × $s$ × $(100\text{ cm})^{-2}$ = 10 poise

The dimensional formula for the coefficient of viscosity is $ML^{-1}T^{-1}$

## 2.0. Stokes law

This law was proposed by the Irish theoretical physicist and mathematician Sir George Gabriel Stokes (1819–1903). Stokes spent his entire career at the University of Cambridge, where he served as the Lucasian Professor of Mathematics from 1849 until his death in 1903. Stokes made seminal contributions to fluid dynamics. He was secretary and then president of the Royal Society. The famous Stokes law relates viscous force $F$ with various factors and is given by

$$F = 6\pi\eta rv \tag{6.3}$$

where $\eta$ is the coefficient of viscosity, $r$ is the radius of a small sphere, and $v$ is the constant velocity with which the body falls. This basic equation is related in studies in fluid dynamics. The above equation was derived in 1845 and is applicable under five conditions

(i) When the body is a small sphere
(ii) Falls though a medium of infinite extent
(iii) The medium is viscous
(iv) The medium in which the body falls remains at rest
(v) The body falls with slow velocity

### 2.1. Derivation of $F = 6\pi\eta rv$

There is no rigorous mathematical derivation for the equation $F = 6\pi\eta rv$. It is derived easily by the method of dimensions.

Let viscous force depend upon $a^{th}$ power of $\eta$, $b^{th}$ power of $r$, and $c^{th}$ power of $v$. Then

$$F \propto \eta^a r^b v^c$$

or
$$F = k\, \eta^a r^b v^c \tag{6.4}$$

where $k$ is the constant of proportionality.

Dimensional equation for force, $F = MLT^{-2}$
Dimensional equation for coefficient of viscosity $\eta = ML^{-1}T^{-1}$
Dimensional equation for radius, $r = [L]$
Dimensional equation for velocity, $v = [LT^{-1}]$

Now setting dimensional formulae of various quantities in Eq. (6.4) we get

$$MLT^{-2} = k\,(ML^{-1}T^{-1})^a\,(L)^b\,(LT^{-1})^c$$
$$= k\, M^a\, L^{-a+b+c}\, T^{-a-c} \tag{6.5}$$

The mass, length and time are fundamental quantities, their dimensions must be the same in both in LHS and RHS. So comparing the dimensions of $M$, $L$ and $T$ we get

$$a = 1 \tag{6.6}$$

$$-a + b + c = 1 \tag{6.7}$$

$$-a - c = -2 \tag{6.8}$$

From Eqs. (6.6) and (6.8)

$$-1 - c = -2 \quad \text{or } c = -1 + 2 = 1$$

Further, substituting values of $a$ and $c$ in Eq. (6.7), we get

$$-1 + b + 1 = 1 \quad \text{or } b = 1 \tag{6.10}$$

Thus

$$a = b = c = 1 \tag{6.11}$$

Substituting the values of $a$, $b$, and $c$ in Eq. (6.4), we get

$$F = k\,\eta r v \tag{6.12}$$

$k$ is the coefficient of proportionality (varies from one situation to other), its value is determined experimentally.

### *2.2. Determination of the magnitude of k – coefficient of proportionality*

Here $k$ is the coefficient of proportionality, as its value varies from one situation to another like the coefficient of thermal conductivity, the coefficient of expansion etc. The value of the constant of proportionality remains the same for all cases. In the method of dimensions the coefficient/constant of proportionality is determined experimentally, it is an established procedure. The same also applies to the viscous force. The value of the universal gravitational constant $G$ is also determined experimentally. George Stokes was the theoretical physicist and Lucasian Professor of Mathematics, so he did not conduct experiments to determine the value of $k$. So Stokes may have definitely given the value of $k$ as $6\pi$, but did not experimentally confirm it. So other scientists might have tried to calculate the value of $k$, or without attempting to calculate the value of $k$, they calculated the value of constant velocity with the theoretical value of $k$ as $6\pi$. The value of the constant of proportionality is regarded as $6\pi$, i.e.

$$k = 6\pi \tag{6.13}$$

Now, how and why the value of $k$ is regarded as $6\pi$? Under the established rules, the value of the coefficient of proportionality must be determined experimentally when the equation is derived experimentally. Then it is just possible that in numerous cases values of constant $k$ are feasible. In the existing literature there is no experimental data which illustrates experimental determination of $k$ as $6\pi$. Was the value of $k$ as $6\pi$ regarded as a postulate or is it a theoretical estimate or experimental determination?

If the value of $k$ is estimated, where are the details of the experiments? When and where were these experiments conducted? How many observations were taken? What are the conditions under which experiments were conducted? Are these experiments (if conducted) repeatable? Further, there is also relative motion between various layers of fluids when bodies rise upwards. For falling bodies the equation for the viscous force is

$$F = k\eta rv = 6\pi\eta rv = 18.8571\ \eta rv \tag{6.14}$$

## 2.3. Constant velocity attained by a falling body

According to the Stokes law, when a body falls through a viscous medium it experiences a viscous force. In addition to this, the body experiences a downward force due to the weight and upthrust in the medium in the upward direction due to the Archimedes principle. The magnitude of the viscous force increases with increase in velocity. A stage comes when the viscous force balances the net force, then the body starts moving with constant velocity, called terminal velocity, $v$. The weight of the small sphere, the upthrust experienced by the sphere in the medium and the viscous drag act in the following way.

(i) The weight of the body acting vertically downwards which is given by Eq. (6.15) as

$$W = mg = D_b Vg \tag{6.15}$$

The weight of the body acts in the downward direction.

(ii) The upthrust experienced by the body in the medium of density $D_m$

$$\begin{aligned} U &= \text{weight of the fluid displaced by the body} \\ &= D_m Vg \end{aligned} \tag{6.16}$$

The upthrust acts in the upward direction when the body is immersed in the fluid.

(iii) The viscous drag force acts in the direction opposite to the one in which the body moves. As the body moves downwards, the viscous force acts in the upward direction.

$$F = 6\pi\eta rv. \tag{6.14}$$

The viscous force acts in the upward direction as the body falls downwards.

### *2.4. The three forces compete*

As the body falls down, its downward speed keeps on increasing due to gravity as motion is accelerated. Consequently, the opposite viscous drag ($F \propto v$), which acts in the upward direction also increases, thus the variation in velocity decreases. The upthrust acts upward. Then a stage comes when the weight of the body is just equal to the sum of upward thrust and upward viscous forces. At this stage there is no net force to accelerate the body. Hence the body starts moving with a constant velocity. When the body attains constant velocity then

$$U + F = W \tag{6.17}$$

Up thrust + Viscous Force = Weight

Total upward force = Total downward force

$$D_m Vg + 6\pi\eta rv = D_b\, Vg \tag{6.18}$$

In the Stokes law we consider a small spherical body (not bodies of different shapes), thus its volume $V$ is

$$V = \frac{4}{3}\pi r^3 \tag{6.19}$$

Now substituting the value of the volume from Eq. (6.19) in Eq. (6.18), we get

$$D_m \frac{4}{3}\pi r^3 g + 6\pi\eta rv = D_b \frac{4}{3}\pi r^3\, g$$

$$6\pi\eta rv = D_b \frac{4}{3}\pi r^3\, g - D_m \frac{4}{3}\pi r^3 g$$

or

$$6\pi\eta rv = (D_b - D_m)\, \frac{4}{3}\pi r^3\, g \tag{6.20}$$

$$v = \frac{2r^2\left(D_b - D_m\right)g}{9v} \tag{6.21}$$

Equation (6.21) can also be used in understanding the motion of bodies in various fluids. For example

(i) If the density of a falling body is the same as that of the medium in which the body falls, i.e.

$$D_m = D_b \tag{6.22}$$

Now Eq. (6.21) becomes

$$v = \frac{2r^2(D_b - D_m)g}{9v} = 0 \tag{6.23}$$

It means if both the body and the medium have the same density then the body does not move. In this case the body just floats in the medium.

(ii) If the density of the falling body is more than that of the medium, it falls, i.e.

$$D_b > D_m$$

Now Eq. (6.21) becomes

$$v = +ve$$

The positive sign indicates that the body falls in the medium. It means if the density of the body is more than that of the medium then the body falls down. When the body falls down then the acceleration due to gravity $g$ is also taken as positive.

(iii) If the density of the falling body is less than that of the medium in which the body rises, i.e.

$$D_b < D_m$$

Now Eq. (6.21) becomes

$$v = -ve$$

The negative sign indicates that the body rises in the medium. It means if the density of the body is less than that of the medium then the body falls upward. When the body rises upward then the acceleration due to gravity $g$ is also taken as negative ($-g$).

Now the Stokes law is regarded as to hold good in rising bodies as well. For falling bodies it holds good under five conditions. Such conditions have not been defined for rising bodies. For complete understanding the Stokes law such conditions have to be defined or ascertained for rising bodies and must be tested experimentally over a wide range. Even in the case of falling bodies the law is not experimentally confirmed over a wide range.

All these deductions are consistent with the Archimedes principle. The reason is that the Archimedes principle explains the phenomena in a similar way. In the Archimedes principle, the resultant weight is calculated using weight and upthrust. This interpretation is general and qualitative in nature.

The Stokes la is applicable under five postulates in the case of falling bodies in media. Thus, the law is applicable in a limited region only. In addition to weight, the upthrust is considered along with the viscous force. Thus, finally the constant velocity of the falling body is calculated as in Eq. (6.21). Now this equation can be used to calculate the distance travelled by the body using the equation of uniform velocity

$$S = vt \tag{6.24}$$

Now Eq. (6.24) becomes

$$S = \frac{2r^2\left(D_b - D_m\right)g}{9v}t \tag{6.25}$$

But this equation is applicable in a limited region, like the Stokes law. Now the value of the coefficient of viscosity is given by

$$\eta = \frac{2r^2\left(D_b - D_m\right)g}{9v} \tag{6.26}$$

Now measuring various values, i.e. $r$, $D_b$, $D_m$, $v$ and $g$, the coefficient of viscosity can be measured. The instruments known as viscometers are used for the purpose, e.g. Ostwald viscometer.

## *2.5. Is Stokes law applicable to rising bodies?*

In the same medium depending upon the density, the bodies fall and rise. For example an aluminium sphere falls in water and cork rises upward. The medium possesses the density and viscosity which affect both rising and falling bodies. For the falling bodies there is considerable experimental and theoretical data. But for the rising bodies there are no experimental observations, so that conclusions may be drawn.

The small spheres fall down in water (or denser fluids) with constant velocity, due to viscous drag of the medium. A similar effect can be observed in water for small rising spheres under suitable conditions. Thus, sensitive experiments must be conducted to understand the Stokes law in the case of rising bodies. It will help not only understanding the Stokes law but also the basic laws of science (fluid dynamics). The body is expected to rise upward under the following conditions:

- (i) The density of the body must be comparable to that of the the medium, so that it may easily attain the constant velocity in the medium
- (ii) When the body is a small sphere
- (iii) Rises through a medium of infinite extent
- (iv) Medium is viscous
- (v) The medium in which the body rises remains at rest
- (vi) The body rises with slow velocity

## 3.0. Measurement of viscosity of fluids by the method of rising bodies

In the existing literature, some methods for measuring the viscosity of fluids involve small spheres falling through the fluids [1–4]. On the same principles, the viscosity of fluids can be determined by using small spheres or balloons rising in the fluids. But mathematical equations and experimental description for rising small spheres are not available in standard references. It is formulated here. The possible reason for lack of information is that the most of the methods for determination of viscosity acquired the mature status in the earlier part of the 19$^{th}$ century. The value of $\eta$, determined by the method of falling spheres, was used in various phenomena and no new method (of rising bodies) has been expedited or investigated.

In 1913–1914, Dr. Millikan utilized the existing values of the coefficient of viscosity of air in determination of the electronic charge [4]. As in this case the medium is a gas, the equation for terminal velocity, i.e. Eq. (6.21), was used in modified form [7]. It happened after the death of the original inventor G. Stokes who had proposed the law in 1845. At that time the viscosity of air was determined by the capillary tube method, the constant deflection method, damping of oscillating cylinders, the method of flow through pipes, etc. Probably, scientists found these methods good enough for the purpose. Consequently, they did not investigate it by the method involving rising small balloons/bubbles. Also at that time such experiments may have been technically and conceptually comparatively more complex. In the case of rising bodies, even it is not confirmed under which conditions, the body moves with constant velocity. Currently, the central theme of discussion is the formulation of mathematical equations and description of experimental methods of measurement of the viscosity of fluids involving rising spheres in fluids. These investigations are important in the sense that the conditions of validity for the rising bodies are yet to be established.

### *3.1. Mathematical background*

The mathematical treatment requires the viscous force (resistive in nature) experienced by small spheres in fluids. This force arises due to the velocity gradient in the system of fluids and spheres. The bodies attain constant velocity due to this force which is given by,

$$F = k\eta rv = 6\pi\eta\, rv \;=\; 18.8571\ \eta rv \tag{6.14}$$

The experimental set up for measurement of the value of $k$ as $6\pi$ (18.857) is not described in the existing literature. Stokes under five assumptions put forth the equation (6.14) in 1845. In 1910, Arnold [4] experimentally justified that it holds good with reasonable accuracy for falling spherical particles of rose metal with a radius of 0.002 cm in water with an accuracy of few tenths of 1%.

Now it can be safely concluded that as the viscous force is applicable in the falling bodies, it must also hold good in the case of the rising bodies. The viscous force opposes the motion of the body, and it acts in the direction opposite to that of the motion of the body. In the same fluid the different bodies rise and fall. If the Stokes law is applicable for the falling bodies under some conditions, the same must hold good for rising bodies also under some conditions. However, in the case of the rising bodies the conditions of applicability may be different. In the existing literature there are no references that such discussion was ever done on the topic. So here our aim is to discuss this aspect theoretically. Let the viscous force in this regard hold well as

$$F = k_r\ \eta rv \tag{6.27}$$

where $k_r$ is the coefficient of proportionality, $r$ is the radius, $v$ is the constant velocity of a small sphere or balloon when it rises upward and $\eta$ is the coefficient of viscosity of fluid. The value of $k$ for small spherical particle is regarded as $6\pi$ (18.8571), but for the falling bodies only in a particular range. Otherwise, the Stokes law is only used in modified form, e.g. in the Millikan oil drop experiment. For the falling bodies Eq. (6.21) was verified for small spheres of rose metal (alloy) with a radius of 0.002 cm by Dr Arnold in 1910 with an accuracy of a few tenths of 1%. The repeatability of experiments is not mentioned. The repeatability of experiments should be observed not only in water but in all fluids and solid bodies. There is no experimental data if rose metal has a radius of 0.003 cm or more, i.e. when its velocity is uniform or variable. Similar is the case of numerous other metals and alloys. So even in the case of the falling bodies the experimental verification is not complete. So the validity of the equation

$$F = k\eta rv = 6\pi\eta\ rv = 18.8571\ \eta rv \qquad (6.27)$$

is used as a postulate even in case of the falling bodies. But the equation equivalent to Eq. (6.27) for the rising bodies is not discussed at all in the existing literature.

### *3.2. A suitable body rising upward in a viscous medium*

The motion of the rising body is considered identically as in the previous case of the falling bodies. When a lighter body is left free from the bottom in the vessel filled with a liquid, then it rises upwards. The weight of the small sphere, the upthrust experienced by the sphere in a medium and the viscous drag due to the medium act in the following way.

(i) The weight of body (due to gravity) acting vertically downwards which is given by Eq. (6.15) as

$$W = mg = D_b Vg \qquad (6.15)$$

The weight of the body acts in the downward direction due to gravitational attraction.

(ii) The upthrust experienced by body (due to the Archimedes principle) in a medium of density $D_m$

= equal to the weight of fluid displaced by the body

$$= D_m Vg \qquad (6.16)$$

It acts in the upward direction due to the Archimedes principle.

(iii) The viscous drag force acts in the direction opposite to the one in which the body moves. In this case the body moves upwards and the viscous force moves in the downward direction

$$F_r = k_r\ \eta rv \qquad (6.27)$$

For the falling bodies the Stokes law determined the value of $k$ as $6\pi$. $F_r$ is the viscous force for the rising bodies, $k_r$ has to be determined. The viscous drag directly depends upon the coefficient of viscosity $\eta$, the radius of the sphere $r$ and constant velocity $v$.

Let us assume that a body of suitable size, having the density comparable to that of the medium, rises upward with constant velocity due to the upthrust. The viscous force and weight act downwards. If the density of the body is comparable to that of the medium then it will attain the constant velocity easily. The small spheres attain constant velocity due to viscous drag under certain conditions. If these conditions are not satisfied then the velocity is variable. In this case motion is expected to be accelerated. It has already been mentioned that the viscous drag acts in the opposite direction, i.e. as the body moves upwards the viscous drag acts in the downward direction. When the body attains constant velocity (under above conditions) then

$$W + F = U \tag{6.28}$$

Weight + Viscous Force = Upthrust

$$D_b Vg + k_r \eta rv = D_m Vg \tag{6.29}$$

The Stokes law considers a small spherical body so that its volume $V$ is

$$V = \frac{4}{3}\pi r^3 \tag{6.19}$$

Now substituting the value of the volume from Eq. (6.19) in Eq. (6.29), we get

$$D_b \frac{4}{3}\pi r^3 g + k_r \eta rv = D_m \frac{4}{3}\pi r^3 g$$

$$k_r \eta rv = D_m \frac{4}{3}\pi r^3 g - D_b \frac{4}{3}\pi r^3 g \tag{6.30}$$

or $$k_r \eta rv = (D_m - D_b)\, \frac{4}{3}\pi r^3 g$$

$$v = \frac{4\pi r^2 \left(D_m - D_b\right) g}{3k_r \eta} \tag{6.31}$$

As in case of the falling bodies, the distance travelled in this case is

$$S = \frac{4\pi r^2 \left(D_m - D_b\right) g}{3k_r \eta}\, t \tag{6.32}$$

Equation (6.31) gives theoretically the value of constant velocity with which the body rises under the effect of the viscous force. Now determining the value of constant velocity $k_r$ can be determined. The magnitude of the coefficient of viscosity is given by

$$\eta = \frac{4\pi r^2\left(D_m - D_b\right)g}{3k_r\,\eta} \tag{6.33}$$

Now with help of Eq. (6.33) the value of the coefficient of viscosity can be measured. The first and foremost condition for measurement of $k_r$ is that it must rise upward with constant velocity. If values of various variables, i.e. $r$, $D_m$, $D_b$, and magnitude of $\eta$, are known then the value of $k_r$ can be measured.

Equation (6.33) can be written as in the form of Eq. (6.27). For this, let us write the equation as

$$F = K6\pi\eta\ rv \tag{6.34}$$

where $K$ is a factor different from unity and non-zero. Here it is assumed that the viscous force for the rising bodies is different from $F = 6\pi\eta\ rv$.

Now the equation equivalent to Eq. (6.29) for the rising bodies can be written as

$$K6\pi\eta\ rv = \frac{4}{3}(D_m - D_b)\ r^3g \tag{6.35}$$

or

$$v = \frac{2r^2(D_m - D_b)g}{K9\eta} \tag{6.36}$$

Further, the distance travelled is given by

$$S = vt = \frac{2r^2(D_m - D_b)g}{K9\eta}t \tag{6.37}$$

Now the value of $K$ can be experimentally confirmed using various values of parameters.

### *3.3. Measurement of viscosity of fluids by equation (6.31)*

In this case, firstly, the conditions are to be determined for which the rising bodies in fluids attain constant velocity (if actually bodies do so). Then the coefficient of proportionality can be determined. Many experiments are required to ascertain this. It is again reiterated: in the existing literature there is no experimental data about the falling bodies which would show how the value of $k$ is determined as $6\pi$

(18.8571). It is just possible that the value of $k$ equal to $6\pi$ (18.8571), may have been determined theoretically, and other data is based upon it, including the electronic charge.

The viscosity of liquids, say water (1000 kg $m^{-3}$), can be determined in the following way. Firstly, the small spheres of a non-hygroscopic plastic have to be fabricated so that their density, i.e. $D_m$, may not vary when the sphere rise upwards. The non-hygroscopic plastic[5] lighter than water may have the composition 93.8% (in weight) of Nylon 12 [i.e., $(C_{12}H_{23}NO)_n$] and 6.2% Santicizer 8 (i.e. $C_{10}H_{15}NO_2S$). The small spheres which have a higher density of 900 kg $m^{-3}$ may acquire constant velocity easily in water. The coefficient of viscosity of glycerine is about 1058 times higher than that of water. So in glycerine the body may attain constant velocity easily.

The small spheres may be allowed to rise upwards in a vessel of suitable dimensions (i.e., depth 2.5 m and radius 0.5 m, say) filled with water and consisting of transparent glass to facilitate experimental observations. Nearly at the bottom of the vessel or tank of water a syringe should be fitted so that non-hygroscopic particles may be smoothly injected in water to rise upwards. Further the small spheres should be cautiously and judiciously injected in water so that water is not set in motion. It can be realized in practice by placing carefully the sphere at the tip of the syringe. Otherwise the movement of water may not allow the small sphere to attain constant velocity easily. Further, the length of the syringe (may be about 0.5 m) should be inside the water so that the small sphere is injected in the middle of the tank or cylinder.

In such experiments if bodies of the shape other than spherical are used then bodies may not attain constant velocity easily. Such bodies may set the water in motion and hence the results may be affected. The existence of the above equations is based upon the constancy of the speed of light. The values of $S$ and $t$ can be measured as in case of falling sphere experiments. Further, for more precise results the distance may be measured by the interferometric method and the time interval by a quartz clock and electronic counters [6] as in the case of determination of $g$ using the following equations.

$$g = \frac{2S}{t^2} \tag{6.38}$$

Further utilizing the same principle, the viscosity of gases (or gas mixtures) may be determined by allowing the balloons filled with lighter gases to rise upward in closed chambers.

## Frequently asked questions

1. **Did George Gabriel Stokes himself experimentally determine the value of $k$ equal to $6\pi$ in equation $F = k\eta rv$?**
   **Answer:** No, Sir George Gabriel Stokes (1819–1903) did not determine the value of $k$ as $6\pi$ experimentally. The law was stated in 1845.
   George Stokes was the theoretical physicist and Lucasian Professor of Mathematics, so he did not conduct experiments to determine value of $k$. There is no evidence in the literature. So Stokes may have definitely given the value of $k$ as $6\pi$, but did not experimentally confirm it. So other scientists might have tried to calculate the value of $k$, or without attempting to experimentally determine the value of $k$, they calculated the value of constant velocity with the theoretical value $k$ as $6\pi$.
2. **Stokes law is the well established law in fluid dynamics. How is the value of the constant of proportionality determined?**
   **Answer:** It is evident from the existing data that the value of $k$ was determined theoretically.
   The Stokes law was stated by the Irish physicist and mathematician Sir George Gabriel Stokes (1819–1903) in 1845. The viscous force in the Stokes law was derived by the method of dimensions, e.g.

$$F = k\eta rv$$

   where $k$ is the coefficient of proportionality. Its value has to be determined experimentally, which is the established norm. But in the existing literature there is no description of experiments for the calculation of $k$. It is assumed that the value of $k$ remains the same for all situations, however some corrections are applied to this.
   If the value of $k$ is experimentally estimated then where are the details of the experiments? If these experiments were conducted then experimental data must be there. What are the conditions under which the experiments were conducted? Are these experiments repeatable? Does the value of $k$ vary from one observation to another? Will the range of applicability of the law $F = 6\pi\eta rv$ increase if the value of $k$ is regarded different from $6\pi$?
   In the existing literature it is mentioned that Dr. Arnold confirmed the Stokes law ($F = k\eta rv = 6\pi\eta\ rv = 18.8571\eta rv$) for rose metal (alloy) of a radius of 0.002 cm ($V = 33.52 \times 10^{-8}$ cc) with an accuracy of few tenths of 1%. It is only calculated for the rose metal in water but not for other metals and fluids. Millikan

also modified the Stokes law [7] in gas while determining the electronic charge.

3 **Is the Stokes law equally applicable to rising bodies as well?**

**Answer:** Yes, theoretically the Stokes law is equally applicable to the rising bodies as well as to the falling bodies. For example, if velocity in Eq. (6.21) is negative then the body moves upward. But the law has no been studied theoretically nor experimentally in the case of the rising bodies. The basis of the Stokes law is the force which arises due to internal friction of fluid (liquid and gas), called the viscous drag or viscous force.

(i) When the body falls down in the medium then the viscous force acts in the backward direction. Taking this as a basis, Stokes put forth an equation under five postulates as

$$F = k\eta rv = 6\pi\eta\ rv = 18.8571\eta rv$$

The value of $k$ (better termed as the coefficient of proportionality) is regarded as equal to $6\pi$.

(ii) A heavier body falls down in a fluid and a lighter body rises upward. When the body rises upward even then the viscous force acts downwards. So in analogy to the falling bodies, under some conditions the body may rise upward with constant velocity. But these conditions are determined experimentally. This situation (rising bodies) is described by equation

$$W + F = U$$

Weight + Viscous Force = Upthrust

The similar equation for the falling bodies is

$$U + F = W$$

So experimental fluid dynamists have a double job in hands. Firstly, the value of $k$ for the falling bodies has to be determined for various materials in different fluids. Secondly, the method for determination of $k_r$ has to be established. Without such specific experiments the understanding of basics of fluid dynamics is incomplete.

## References

1. Rosenberg, R.M., *Principle of Physical Chemistry* (Oxford University Press, Oxford), pp. 508–511 (1st edition) (1977).
2. Barrante, J.R. *Physical Chemistry for Life Sciences* (Prentice-Hall Inc., New Jersey) pp. 292–297 (Ist Edition) (1977).
3. Arya, A.P., *Introductory College Physics* (Macmillan Publishing Company, New York) pp. 253–254 (1st Edition) (1979).
4. Millikan, R.A, *The Electron* (University of Chicago Press, Chicago) pp. 93–96 (2nd edition, 10th impression).
5. Bizzeti, P.G., Bizzeti-Sona, A.M., Fazzini, T., Perego, A., and Taccetti, N., Phy. Rev. Lett., 62, 2901–2904 (1989).
6. Encyclopeadia of Science and Technology, No. 5 (MacGraw Hill Publishing Company, New York), pp. 482–483 (1987).
7. Millikan, R.A., The Physical Review, Vol. II, No. 2, pp. 109–143 (1913).

# 7

# Limitation of Existing Theories and an Alternate Theory of Rising, Falling and Floating Bodies

**First glimpse**

- ✓ Here the natural motion of bodies (fall or rise without external influence) is considered. The bodies fall under the influence of gravity.
- ✓ In existing physics the The Archimedes principle and Stokes law are used to explain the motion of rising, falling and floating bodies.
- ✓ the Archimedes principle states that if $D_b > D_m$ the body rises, at $D_b < D_m$ the body falls and if $D_m = D_b$ then the body floats. It is a qualitative explanation only.
- ✓ The principle does not correctly points out 'how much distance' is travelled (fallen or risen) by the body in time *t,* i.e. 1 s, 4 s or 9 s, etc.
- ✓ According to the Archimedes principle bodies of aluminium (2700 kg/m$^3$) of mass 1 mg (spherical in shape) and 1000 kg (flat in shape) must fall through a distance of 3.085 m in 1 s. But this prediction is not confirmed yet, but regarded as true.
- ✓ The principle only takes in account $D_m$ and $D_b$, and neglects the mass, shape and angle at which body is dropped, the magnitude and characteristics of the motion of the medium and convectional currents, etc.

- ✓ R. Piazza observed anomalous observations to the The Archimedes principle in sensitive experiments – heavy particles of gold floated over the surface of a lighter medium.
- ✓ The Stokes law is applicable in a very narrow range, i.e. Arnold confirmed it for bodies of rose metal of a radius of 0.002 cm with some accuracy, otherwise it is not valid.
- ✓ The drag force is not applicable to natural motion.
- ✓ Thus, an alternate theory on rising, falling and floating is developed taking all factors into account (i.e. mass, shape and angle at which the body is dropped, the magnitude and characteristics of motion of the medium and convectional currents, etc.)
- ✓ The newly discussed terms are the hidden ratio, the falling factor and the rising factor.

## 1.0. Relevant terms

***Buoyancy or upthrust***

**Fluid:** The fluid is the common name for liquids and gases, e.g. air, water, hydrogen, glycerine, etc. the Archimedes principle is equally applicable to all fluids. The magnitude of the upthrust is greater in the case of liquids than gases, as the density of liquids is more than that of gases.

**Buoyancy or buoyant force or upthrust:** Buoyancy (also known as the buoyant force or upthrust) is the upward force exerted on an object by the fluid when it is wholly or partly immersed in a fluid. It is a vector quantity having both magnitude and direction. Buoyancy acts upward for the kind of situations encountered in everyday experience.

Upthrust is a force that pushes an object upwards when it is in a fluid (a liquid or gas). Upthrust is also called buoyancy. Due to this the object appears to be lighter in a liquid or weigh less. In the SI system the unit of buoyancy is the newton [*N*].

***Mass***

**Mass** is a measure of the amount of material in an object. Mass does not change with body's position, movement or alteration of its shape, unless material is added or removed (the results of special theory of relativity are not taken in account). The mass is a fundamental property of an object, a numerical measure of its inertia and a fundamental measure of the amount of matter in the object. Mass is measured with a physical balance. The units of mass are kilogram in the SI system, and gram in the CGS system.

Mass and weight are often misunderstood terms in various branches of physics. The fundamental relation between the mass and acceleration is defined by Newton's second law and can be expressed as

$$F = ma \tag{7.1}$$

where

$F$ = force (N)
$m$ = mass (kg)
$a$ = acceleration ($m/s^2$)

$$a = \frac{(v-u)}{t} \tag{7.2}$$

where $v$ is the terminal velocity, $u$ the initial velocity and $t$ is time.

***Weight***

**Weight** is the gravitational force acting on a body mass. The weight of the body obtained from Newton's second law of motion by replacing $g$ by $a$ in $F = ma$ is

$$W = m\,g \tag{7.3}$$

where

$W$ = weight, measured in Newton (N). The gravitational unit of weight is 1 kg wt (9.8 N)
$m$ = mass, measured in kilogram (kg)
$g$ = acceleration of gravity, it has units ($m/s^2$) and dimensions of acceleration ($M^0LT^{-2}$)

Weight is measured with a spring balance.

In the SI system the mass unit is the kg and since the weight is a force – the weight unit is the Newton (N).

A car's mass is 1 000 kg. The weight can be calculated:

$$W = mg = (1000\text{ kg})(9.807\text{ m/s}^2) \tag{7.4}$$
$$= 9807\ N = 9.807\text{ kN}$$

***Density***

**Density** is defined as object's mass per unit volume.

The density can be expressed as

$$D = \frac{M}{V} \tag{7.5}$$

where

$$D = \text{density}\left(\frac{kg}{m^3}\right)$$

$M$ = mass (kg)
$V$ = volume ($m^3$)

The SI units for density are $\left(\frac{kg}{m^3}\right)$.

The densities of air, water and steel are 0.001293 kg/$m^3$, 1000 kg/$m^3$ and 7800 kg/$m^3$ respectively. In general understanding the heavier body has greater density than the lighter body. It is evident from the existing literature, i.e. book The Works of Archimedes by T.L. Heath. In the book Archimedes states that the heavier body falls in the medium and the lighter body rises.

According to the T.L. Heath's book [1] The Works of Archimedes (On floating bodies, Book I) on page 253, we have Preposition 7 as "A solid heavier than a fluid will, if placed in it, descend to the bottom of the fluid, and solid will, when weighed in the fluid, be lighter than its true weight by the weight of the fluid displaced."

**Thrust:** The total force exerted by a liquid on any surface in contact with it, is called the thrust of the liquid. It is denoted by $F$.

**Pressure:** The thrust exerted by a liquid (at rest) per unit area $A$, then pressure is given by

$$P = \frac{F}{A} \tag{7.6}$$

The unit of pressure in the CGS system is dyne/$cm^2$ and N/$m^2$ in the SI system. 1 N/$m^2$ is also called Pascal (Pa)

1 Pa = 10 dyne/$cm^2$

The pressure is a scalar quality

### *1.1. Archimedes' principle*

In antiquity science is developed in different regions in different ways. Every development is significant and noticeable. In Greece Aristotle (384–322 BC) and Archimedes (287–212 BC) did significant work in science which is taught even after thousands of years. Although

Aristotle's work has been considerably modified or overhauled, yet the principle of Archimedes still stands in the original form in spite of the fact being 2265 years old [2]. Today, technological advancements are touching the sky compared to the days of Archimedes. It is true that the Archimedes principle has passed modern day tests, especially regarding the measurement of density. But it did not end the investigations in the scientific world. The principle is yet to be checked on many fronts. But the mathematical equations based upon the Archimedes principle became derivable in 1685 when Newton published the Principia [3] and defined the acceleration due to gravity. How did scientists take the principle granted for 1935 years without mathematical equations? It is just someone's guess.

However, recently the principle has been generalized in two ways.

(i) It is pertinent to mention that in 2012 for the first time in history of science some sensitive experiments were reported by a team of Italian scientists headed by Roberto Piazza and others[4]. The results were not consistent with the Archimedes principle, e.g. the Archimedes principle states that:

A body heavier than the medium must sink in the medium whereas experimentally it was found that gold particles (19.2 gm/cc) of exceptionally small in size float in a lighter medium. The density, mass and volume of gold particle do not change in the medium. The gold particles should have sunk in the lighter medium.

This is the first results reported in sensitive experiments which was found inconsistent with the 2265 years old Archimedes principle. The aim is that such cases must be explored to all significant cases. They have generalized the principle by adding another term to the original form of the Archimedes principle in paper which was published by the Royal Society of Chemistry, London England. Similar experiments were also reported by D. Heras in Nature Scientific Reports published by the Nature Publishing Group [5].

(ii) Also in 1997 the author has reported similar first stage experiments that a umbrella-shape heavier body may float in water. Earlier the author generalized the principle on the basis of a theoretical study, i.e. in one case the principle predicts that the volume of the medium filled in the balloon becomes indeterminate, i.e.

$$V = \frac{0}{0}$$

In this case, the principle is generalized in the form of a proportionality, i.e upthrust is proportional to the weight of the fluid displaced

$$U = f\,DVg \tag{7.7}$$

where $f$ is the coefficient of proportionality. When the oldest established principle is generalized, we get exact volume,

$$V = V \tag{7.8}$$

The generalization of the principle is justified.

## 2.0. The Archimedes principle in rising, falling and floating bodies

According to the anecdote Archimedes stated the principle for the measurement of the density of bodies, king's crown to be more specific. It is pertinent to mention here that Archimedes enunciated the principle for determination of densities and at later stage its applications were extended to phenomena of rising, falling and floating bodies. In the literature there are no hints that Archimedes ever applied the principle to rising, falling and floating bodies, as cited above in section 1.0. But even now quantitative experiments over a wide range of parameters have not been conducted yet. Based upon the Archimedes principle, the quantitative mathematical equations have been developed for rising and falling bodies. The results of these experiments would be significant in this regard. Now these equations can be experimentally tested in sensitive experiments. Thus, according to the Archimedes principle if $D_m$ and $D_b$ are the same, then all bodies, irrespective of mass and shape (defined earlier), should fall or rise with the precisely constant acceleration. But this fundamental prediction has not been experimentally confirmed as yet.

Anyhow, let us see how we understand the phenomena of rising, falling and floating bodies now. Thus, equations can be formulated with the help of the law of gravitation using the concepts of weight and upthrust. The same equation of the resultant weight is used to explain the phenomena of rising, falling and floating bodies. From the resultant weight, the resultant acceleration can be calculated, hence the distance travelled in time $t$. This attempt is made to illustrate the distance travelled by the body in time $t$.

## *2.1. Qualitative explanation*

Consider a body of mass, $m$ and volume $V$, then its weight is

$W$ = mass × acceleration due to gravity

$$= mg \tag{7.3}$$

Further, mass can be written in terms of density

$$m = DV \tag{7.5}$$

Substituting Eq. (7.5) in Eq. (7.3) we get

$$W = D_b Vg \tag{7.9}$$

When the body is immersed in a fluid (medium) it experiences upthrust and its weight decreases, which is everyday's observation in daily life.

Upthrust, $U$ = weight of fluid displaced

= mass of fluid displaced, $g$

= density of fluid; volume of body $g$

$$= D_m Vg \tag{7.10}$$

The weight of the body is the gravitational pull towards the Earth. When a body is immersed in the fluid its weight decreases.

The resultant weight is the difference between weight and upthrust [6].

$$W_R = W–U = (D_b–D_m)Vg$$

$$= (1-\frac{D_m}{D_b})VD_b g \tag{7.11}$$

Now we have equations for force and weight.

$$F = ma$$

The equation for weight also follows from here (replacing acceleration by acceleration due to gravity),

$$W = mg$$

$$\text{or } g = \frac{W}{m}$$

Thus dividing Eq. (7.11) with the mass of the body, $m = D_b V$,

$$a_R = (1 - \frac{D_m}{D_b})g$$

Thus, the resultant weight (downward gravitational pull) is a combination of two factors, i.e. weight and buoyant force or upthrust

(i) the weight acts downwards
(ii) the upthrust acts upwards

Both are vectors quantities and hence added algebraically. Both forces compete with each other.

(a) If weight (downward gravitational pull) is more than up thrust or buoyant force experienced by the body in the fluid then the body falls down. Hence the body falls down if

$$\text{weight} > \text{upthrust}$$
$$D_b Vg > D_m Vg$$
$$D_b > D_m$$

For example a body of aluminium falls in water as the density of aluminium is 2700 kg/m$^3$ which is more than that of water, 1000 kg/m$^3$. In this case the resultant weight is positive and so is the resultant acceleration. For example

The resultant weight, $W_R = (D_b - D_m)Vg$

$$W_R = (D_b - D_m)Vg = (2700 - 1000)Vg = 1700Vg \tag{7.12}$$

In this case the resultant downward acceleration is positive.

(b) If upthrust is more than weight then the body rises upward

$$\text{upthrust} > \text{weight}$$
$$D_m Vg > D_b Vg$$
$$D_m > D_b$$

In this case, the resultant downward acceleration is negative. For example, a cork body rises in water as the density of cork is 240 kg/m$^3$ which is less than that of water, 1000 kg/m$^3$. In this case the resultant weight is negative, i.e. acceleration is negative. When the body moves upward its acceleration is taken negative by convention.

$$W_R = (D_b - D_m)Vg = (240 - 1000)Vg = -860\ Vg \tag{7.13}$$

The negative sign simply indicates that the direction and the magnitude of resultant weight is 860 $Vg$.

(c) If upthrust and weight are both equal, then the body floats.

$$\text{weight} = \text{upthrust}$$
$$D_b Vg = D_m Vg$$
$$D_b = D_m \tag{7.14}$$

A balloon floats in air as the density of the balloon is equal to that of air. In this case both the resultant weight and resultant acceleration are zero. When acceleration is zero, then the body is at rest or moves with uniform velocity. Here scientists prefer the second possibility only, in which the body remains at rest.

## *2.2. Quantitative explanation*

The above explanation simply implies that if the density of the body is more than that of the medium the body falls down. It is similar to saying that a bus is moving. But none of two are complete statements. The complete statement would be: with what speed is the bus moving, i.e. how much distance it will travel in time $t$? What is the direction of the bus? Likewise, one should be able to find out from the explanation the distance of the fall of the aluminium body water in time $t$. That would be complete information. Simply saying that a body of aluminium falls in water as its density is more is not complete or quantitative statement.

Similar arguments are true for rising bodies as well. But the following query is not explained. Let a body of cork of mass 10 gm rises upward in water, how far will it travel in water in 4 s? In Eqs. (7.11) the resultant weight is related with density and volume. The physical quantity distance is not related with time in the equations. So to explain the falling and rising bodies quantitatively, the relation between displacement and time is needed. Thus equations must be rewritten accordingly. In the existing physics the distance and time are related with the help of constant acceleration (rate of change of velocity). Such kinematic equations are well known in basic physics.

The falling body of aluminium can be understood quantitatively if we can calculate the distance travelled by the body in certain time. The distance is related with time with the help of kinematic equations

$$S = ut + \frac{1}{2}at^2 \tag{7.15}$$

$$S_{nth} = ut + \frac{1}{2}a(2n-1) \tag{7.16}$$

The other kinematic equations relating various parameters are

$$v = u + at \quad (7.17)$$

$$v^2 - u^2 = 2as \quad (7.18)$$

where $S$ is the distance travelled in time $t$, $v$ is terminal velocity, $u$ is initial velocity, $a$ is acceleration. $S_{nth}$ is the distance travelled in $n^{th}$ second. The Eqs.(7.15–7.18) are meant for the uniformly accelerated motion. Equation (7.18) is independent of time and Eq. (7.17) independent of distance. If the body falls under the influence of gravity, then acceleration can be replaced by $g$ in Eqs. (7.15–7.18). If the body moves with uniform velocity then

$$S = Vt \quad (7.19)$$

**Constraints on Eqs.** (7.15–7.18). These equations are derived under the condition that motion is accelerated, i.e. the velocity of the body increases regularly in regular intervals of time. The variable velocity must be such that acceleration is constant. If velocity is uniform then acceleration is zero.

**Constraint on Eq. (7.19).** Equation (7.19) is meant for uniform velocity. It is only applicable if velocity is uniform, not variable. Thus, equations must be applied under the conditions under which they were derived. When the body falls down, it falls with constant acceleration hence Eq. (7.19) is not applicable at all.

### *2.3. Importance of resultant weight*

The distances travelled are expressed in terms of acceleration. In this case, the acceleration is calculated from the resultant weight. The method is very simple as the acceleration is calculated from the force

$$m = \frac{F}{a} \quad (7.1)$$

Likewise, the resultant weight $[W_R = (D_b - D_m)Vg]$ is another form of weight ($W = mg$) which is derivable from $F = ma$.

The resultant weight, $W_R = W - U = (D_b - D_m)Vg$ (7.11)

Also, in terms of mass ($m = D_bV$),

The resultant weight, $W_R = (1 - \frac{D_m}{D_m})D_bVg$ (7.11)

Further the equation in terms of the mass of the body ($m = D_bV$),

The resultant weight $= (1-\frac{D_m}{D_b})mg$ (7.11)

### *2.4. The resultant acceleration follows from the resultant weight*

Now we have the equations for force and weight.

$$F = ma \quad (7.1)$$

The equation for weight also follows from here (replacing acceleration by the acceleration due to gravity),

$$W = mg \quad (7.3)$$

or $$g = \frac{W}{m}$$

Thus dividing Eq. (7.11) with the mass of the body, $m = D_b V$,

$$a_R = (1-\frac{D_m}{D_b})g \quad (7.20)$$

Thus the resultant acceleration due to gravity decreases by factor $(1-\frac{D_m}{D_b})g$. Therefore, in the medium the weight and resultant acceleration decrease due to buoyant force or upthrust. When the density of the body is equal to that of the medium ($D_m = D_b$)

$$\text{The resultant weight, } W_R = (1-\frac{D_m}{D_m})D_b Vg$$

$$= 0 \quad (7.21)$$

$$\text{The resultant acceleration} = a_R = 0 \quad (7.22)$$

Thus, the body neither falls down nor rises upward and hence floats in medium.

Therefore, the body floats in the medium. This situation is like the calculation of terminal velocity. Consequently, when acceleration $a = 0$, then the body either remains at rest or moves with uniform velocity. Scientists prefer to explain it that the body remains at rest.

### *2.5. In vacuum*

Thus from the equation of resultant weight (weight in medium) the resultant acceleration due to gravity ($g_R$) can be calculated by dividing resultant weight by the mass of the body.

The resultant acceleration $g_R$ is given by Eq. (7.20). If the density of the medium is zero i.e. in the case in vacuum, the resultant acceleration due to gravity is

$$a_R = (1 - \frac{D_m}{D_b})g \tag{7.20}$$

If $D_m$ is zero then

$a_R = g =$ acceleration due to gravity in vacuum

It is the correct result. Thus, bodies fall down with reduced acceleration in fluids and with maximum acceleration in vacuum. As the weight of the body decreases in fluids, acceleration also decreases which is described by Eq. (7.20).

Niebauer [7] has confirmed that bodies fall with maximum acceleration $g$ in vacuum.

In vacuum $D_m = 0$, $g_R = g$

Hence in this case ($D_m = 0$) Galileo's deduction that all bodies fall equal distances in equal intervals of time, is precisely true.

In media, the acceleration due to gravity decreases due to the buoyant force or upthrust.

Thus, the body falls with maximum acceleration due to gravity $g$ (9.8 m/s$^2$) in vacuum only. The body falls down as it is attracted due to the inverse square law of gravity with constant acceleration i.e. velocity increases in every second. As bodies fall down the force of attraction also increases. The acceleration increases as the density of the medium decreases.

## 3.0. Archimedes' Principle and quantitative displacements of bodies

Generally, the displacement is given by the kinematic equation when the body falls under the influence of gravity.

$$S = ut + \frac{1}{2}gt^2 \tag{7.15}$$

If the body starts from the rest $u = 0$

$$S = \frac{1}{2}gt^2 \tag{7.23}$$

The bodies fall with constant acceleration in vacuum and fluids as $D_m$ and $D_b$ remain the same. The acceleration in fluids is given by Eq. (7.20) which is based upon the Archimedes principle.

Now replacing $g$ by $g_R$

$$S = ut + \frac{1}{2}(1 - \frac{D_m}{D_b})gt^2$$

If the body starts from the rest, i.e. $u = 0$

$$S = \frac{1}{2}(1 - \frac{D_m}{D_b})gt^2 \tag{7.24}$$

The body falls down obeying the inverse square law of gravitation and in vacuum the acceleration is $g$. Conceptually and mathematically Eq. (7.24) is applicable if the resultant acceleration is constant. Also, according to the Archimedes principle, the resultant downward acceleration ($g_R$) remains same as long as the density of the body ($D_b$) and the density of the medium ($D_m$) are the same. The resultant acceleration is independent of all others factors such as the mass and shape of the body.

But theoretically according to the principle, i.e. Eq. (7.11) the bodies (of different masses and shapes, lighter than medium) also rise with constant acceleration in fluids. The acceleration of the body must be same in a beaker or a bucket or tank of water. Further bodies rise upwards against the inverse square law of gravity (which attracts bodies downwards).

In this case the resultant downward acceleration is negative. What is the reason for upward thrust? Is it the decrease in the gravitational force of attraction? It is just due to the dominant contribution of upthrust or buoyant force. But this prediction is not quantitatively tested at all.

### *3.1. Distance travelled by falling bodies in fluids*

The distance through which the body falls is determined by Eq. (7.24).

Thus distance is calculated from equations

$$\text{Resultant upthrust, } W_R = (D_b - D_m)Vg \tag{7.11}$$

$$\text{Resultant upward acceleration, } g_R = (1 - \frac{D_b}{D_m})g \tag{7.20}$$

Thus the body falls down if resultant the weight is positive, if weight dominates upthrust. In this case, the resultant downward acceleration $a_R$ is also positive. Let us calculate the distances travelled by various bodies in different times. It is only the way to judge the principle quantitatively.

(i) For a body of aluminium (2700 kg/m$^3$) of any mass and shape in water 1000 kg/m$^3$

$$g_R = (1-\frac{D_m}{D_b})g = (1-\frac{1000}{2700})9.8 = 6.170 \text{ m/s}^2 \quad (7.25)$$

$$S = \frac{1}{2}(1-\frac{D_m}{D_b})gt^2 = \frac{1}{2}(6.170 \times 9) = 27.765 \text{ m} \quad (7.26)$$

(ii) For a body of steel (7800 kg/m$^3$) of any mass and shape in water 1000 kg/m$^3$

$$g_R = (1-\frac{D_m}{D_b})g = (1-\frac{1000}{7800})g = 8.543 \text{ m/s}^2 \quad (7.27)$$

$$S = \frac{1}{2}(1-\frac{D_m}{D_b})gt^2 = \frac{1}{2}(8.866 \times 9) = 39.897 \text{ m} \quad (7.28)$$

(iii) For a body of mercury (13 600 kg/m$^3$) in water (1000 kg/ m$^3$)

$$g_R = (1-\frac{D_m}{D_b})g = (1-\frac{1000}{13,600})g = 9.079 \text{ m/s}^2 \quad (7.29)$$

$$S = \frac{1}{2}(1-\frac{D_m}{D_b})gt^2 = \frac{1}{2}(9.079 \times 9) = 40.85\text{m} \quad (7.30)$$

(iv) For a body of gold (19200 kg/m$^3$) of any mass and shape in water 1000 kg/m$^3$

$$g_R = (1-\frac{D_m}{D_b})g = (1-\frac{1000}{19,200})g = 9.289 \text{ m/s}^2 \quad (7.31)$$

$$S = \frac{1}{2}(1-\frac{D_m}{D_b})gt^2 = \frac{1}{2}(9.289)9 = 41.8005 \text{ m} \quad (7.32)$$

(v) For a body of platinum(21600 kg/m$^3$) of any mass and shape in water 1000 kg/m$^3$

$$g_R = (1-\frac{D_m}{D_b})g = (1-\frac{1000}{21,500})g = 9.344 \text{ m/s}^2 \quad (7.33)$$

**Table I.** The resultant downward accelerations and distances travelled by various bodies in water (1000 kg/m³) in 3 s according to the Archimedes principle

| Sr No. | Density of body | Resultant acceleration $g_R = (1 - \frac{D_w}{D_b})g$ | Distance travelled in water $S = \frac{\left(1 - \frac{D_m}{D_b}\right) g t^2}{2}$ | Resultant acceleration dependent on upon $D_m$ and $D_b$ only | Resultant acceleration is independent on all other factors |
|---|---|---|---|---|---|
| 1 | Aluminum (2700 kg/m³) | 6.170 m/s² | 27.765m | yes | yes |
| 2 | Steel (7800 kg/m³) | 8.543 m/s² | 38.44m | yes | yes |
| 3 | Silver (10500 kg/m³) | 8.866 m/s² | 39.897 m | yes | yes |
| 4 | Mercury (13600 kg/m³) | 9.0794 m/s² | 40.8573m | yes | yes |
| 5 | Gold (19200 kg/m³) | 9.2895 m/s² | 41.803m/s | yes | yes |
| 6 | Platinum (21 500 kg/m³) | 9.344 m/s² | 42.048 m | yes | yes |

$$S = \frac{1}{2}(1 - \frac{D_m}{D_b})gt^2 = \frac{1}{2}(9.344 \times 9) = 42.048 \text{ m} \quad (7.34)$$

This equations can also be written for different bodies in different media. The various values of the resultant acceleration and distances travelled in various fluids are shown in Table I.

### *3.2. Motion of bodies in mercury (13 600 kg/m³)*

For a body of gold (19 200 kg/m³) of any mass and shape in mercury(13 600 kg/m³)

$$g_R = (1 - \frac{D_m}{D_b})g = (1 - \frac{13600}{19200})g = 2.8583 \text{ m/s}^2 \quad (7.35)$$

$$S = \frac{1}{2}(1 - \frac{D_m}{D_b})gt^2 = \frac{1}{2} \; 6.9416 \text{ m/s}^2 \times 9\text{s}^2 = 12.86 \text{ m} \quad (7.36)$$

For a body of platinum (21 500kg/m³) of any mass and shape in mercury 13 600kg/m³

$$g_R = (1 - \frac{D_m}{D_b})g = (1 - \frac{13600}{21500})g = 3.6009 \text{ m/s}^2 \quad (7.37)$$

$$S = \frac{1}{2}(1 - \frac{D_m}{D_b})gt^2 = \frac{1}{2} \; 3.6009\text{m/s}^2 \times 9 \text{ s}^2 = 16.204 \text{ m} \quad (7.38)$$

### 3.3. Experiments in glycerine (1260 kg/m$^3$)

The Archimedes principle is meant for fluids (liquids and gases) of different densities, coefficient of viscosities and other characteristics. The density of glycerine is 1260 kg/m$^3$ and its coefficient of viscosity is 10.69 poise, whereas the density of water is 1000 kg/m$^3$ and the coefficient of viscosity is 0.0101 poise. According to the Archimedes principle the viscosity of the medium has no role to play in any way. Thus,

$$\text{Coefficient of viscosity of glycerine} = \\ =1058 \text{ times coefficient of viscosity of water} \quad (7.39)$$

$$\text{Density of glycerine} = 1.26 \text{ times density of water} \quad (7.40)$$

But the Archimedes principle only takes into account the density of fluids; in the case of glycerine the density is nearly equal to that of water, whereas the coefficient of viscosity of glycerine is 1058 times than that of water. Does the exceptionally large value of the coefficient of viscosity play any significant role in case of rising, falling and floating bodies? The viscous force acts in the opposite direction to the motion of bodies, so it may affect the status of the motion of bodies. In falling bodies the Stokes law is confirmed in case of a small sphere of rose metal with a radius of 0.002 cm in water with the accuracy of few tenths of one percent. It is due to the viscous force. If upward and downward speeds are found different then these would be due to the viscosity of fluids only. The reason is that the Archimedes principle takes in account $D_b$ and $D_m$ only. The same is true for floating bodies. These aspects can be precisely studied in specific experiments.

### 3.4. Experiments in mercury (13 600 kg/m$^3$)

As already mentioned, the density of water is 1000 kg/m$^3$ and that of mercury 13 600 kg/m$^3$.

The coefficient of viscosity of water is 0.0101 poise and that of mercury 0.0155 poise. Will the exceptionally large density of mercury affect the results? It can be experimentally observed. The distances through which the body falls in mercury are given in Eqs. (7.35–7.38) and through which the body rises in mercury are shown in Table II. It can be theoretically concluded that along with density, the fluidity or mobility of the medium also plays a significant role in the motion of bodies. Just like Eqs. (7.39–7.40), we can derive an equation for the fluidity or mobility (the term which is not widely studied),

**Table II.** The resultant upward accelerations and distances travelled by various bodies in 3 s in water (1000 kg/m³) and mercury (13 600 kg/m³) according to the Archimedes principle. The distances travelled are independent of the shape and mass of body

| Sr No. | Body of any shape and mass | Resultant acceleration in water $gR = (1 - \frac{D_w}{D_b})g$ | Distance travelled in water $S = \frac{\left(1 - \frac{D_w}{D_b}\right) gt^2}{2}$ | Resultant acceleration in mercury $gR = (1 - \frac{D_m}{D_b})g$ | Distance travelled in mercury $S = \frac{\left(1 - \frac{D_m}{D_b}\right) gt^2}{2}$ |
|---|---|---|---|---|---|
| 1 | Cork (240 kg/m³) | 29.4 m/s² | 139.65 m | 545.533 m/s² | 2,454.89 m |
| 2 | Willow wood 420 kg/m³) | 13.533 m/s² | 60.9 m | 317.33 m/s² | 1,383.9 m |
| 3 | Wood elm (600 kg/m³) | 6.533 m/s² | 29.4 m | 212.333 m/s² | 955.498 m |
| 4 | Wax (900 kg/m³) | 1.0888 m/s² | 5.33 m | 138.288 m/s² | 622.3 m |

Fluidity of mercury = 13.6 Fluidity of water (7.41)

or Density of mercury = $z$ Density of water (7.42)

$z$ has to be determined experimentally.

The experiments in mercury would be significant and need to be conducted with utmost accuracy. Although all experiments in various liquids with various bodies would be significant, the experiments in mercury and glycerine would be of paramount importance. The strange aspect in this regard is that no experiments have been conducted specifically. Thus, the study is not quantitatively complete.

### *3.5 Distance travelled by rising bodies in fluids*

The main aim of the discussion is to develop mathematical equations so that the distance travelled by the body in time $t$ is calculated. It will then be attempted to confirm predictions experimentally. If the predictions are confirmed then the Archimedes principle would be regarded as quantitatively true in such cases. Currently, the equations relating the distance with time $t$ involve acceleration. The resultant downward acceleration can be calculated from the resultant weight by diving it by the mass.

$$\text{Resultant weight, } W_R = (D_b - D_m)Vg$$

$$= (1 - \frac{D_m}{D_b})D_b Vg \tag{7.11}$$

or $$\text{Resultant downward acceleration} = (1 - \frac{D_m}{D_b})g \tag{7.12}$$

$$S = \frac{1}{2}(\frac{D_m}{D_b} - 1)gt^2 \tag{7.24}$$

Similarly, the equation for the resultant upward thrust is given by

$$\text{Resultant upward thrust, } U_R = (D_m - D_b)Vg = (\frac{D_m}{D_b} - 1)VD_b g \tag{7.43}$$

Analogous to Eq. (7.20), the resultant upward acceleration can be calculated by dividing $m = D_b V$

$$\text{Resultant upward acceleration, } h_R = (\frac{D_m}{D_b} - 1)g \tag{7.44}$$

(i) For a body of cork (240 kg/m³) of any shape and mass in water (1000 kg/m³)

$$\text{Resultant upward acceleration} = (\frac{D_m}{D_b} - 1)g = (\frac{1000}{240} - 1)g$$

$$= 31.033 \text{ m/s}^2 \tag{7.45}$$

The resultant downward acceleration is negative (31.033 m/s²), the negative sign indicates that the body rises upward in the medium. When the body is thrown into a fluid (air, say) the value of $g$ is regarded as negative ($-g$) in existing physics. The magnitude of acceleration is 31.03 m/s².

$$S = \frac{1}{2}(\frac{D_m}{D_b} - 1)gt^2 = \frac{1}{2}(31.03 \times 9) = 139.65 \text{ m} \tag{7.46}$$

(ii) For a body of willow wood (420 kg/m³) of any shape and mass in water (1000 kg/m³)

or $$\text{Resultant upward acceleration} = (\frac{D_m}{D_b} - 1)g$$

$$= (\frac{1000}{420} - 1)g = 13.533 \text{ m/s}^2 \tag{7.47}$$

The resultant downward acceleration is negative, the negative sign indicates that the body rises upward in the medium. The magnitude of acceleration is 13.533 m/s².

$$S = \frac{1}{2}(\frac{D_m}{D_b} - 1)gt^2 = \frac{1}{2}(13.533 \times 9) = 60.9 \text{ m} \tag{7.48}$$

(iii) For a body of wood elm (600 kg/m³) of any shape and mass in water (1000 kg/m³)

Resultant upward acceleration $= (\frac{D_m}{D_b} - 1)g = (\frac{1000}{600} - 1)g =$

$$= 6.533 \text{ m} \tag{7.49}$$

The resultant downward acceleration is negative, the negative sign indicates that body rises upward in the medium.

$$S = \frac{1}{2}(\frac{D_m}{D_b} - 1)gt^2 = \frac{1}{2}(6.533 \times 9) = 29.4 \ m \tag{7.50}$$

(vi) For a body of wax (900 kg/m³) of any shape and mass in water (1000 kg/m³)

Resultant upward acceleration $= (\frac{D_m}{D_b} - 1)g = (\frac{1000}{900} - 1)g = -1.0888$ m/s²

The resultant downward acceleration is negative (1.0888 m/s²), the negative sign indicates that the body rises upward in the medium. The magnitude of acceleration is 1.0888 m/s²

$$S = \frac{1}{2}(\frac{D_m}{D_b} - 1)gt^2 = \frac{1}{2}(1.0888 \times 9) = 4.9 \ m$$

Similarly, equations can be written for various bodies in mercury (13 600 kg/m³).

Theoretically, the Archimedes principle predicts that a body of cork (240 kg/m³), irrespective of mass and shape, should rise through 139.65 m in 3 s in water. In above calculations the time is regarded as 3 s, this interval can be regarded as smaller than 3s for simplicity. The same body of cork must rise upwards 2.499 m in 3 s in mercury. There is no experimental evidence in the existing literature that such experiments have ever been conducted. Also, there are no hints as to when such equations were formulated. The media of different dimensions should be considered to draw final deductions; as size of medium is likely to be very significant.

In the equation concerning rising and falling bodies the acceleration due to gravity $g$ (9.8 m/s²) appears in mathematical equations. So the distance travelled (fallen or risen), is the multiple of acceleration due to gravity, $g$. It is justified in falling bodies as the bodies fall with the same acceleration in vacuum (9.8 m/s²), but no observations are possible in the case of rising bodies. The upward acceleration is also the multiple of $g$ (9.8 m/s²). Thus such experiments are absolutely necessary regarding rising bodies, as the force of attraction between the body and the Earth increases according to the inverse square law. The generalized theory is only meant for the natural motion of bodies.

**Constraints on Eqs. (7.20–7.24).** Equations (7.20–7.24) are only applicable if the resultant downward acceleration is precisely constant. The resultant acceleration will be constant if the medium has a sufficiently large magnitude (and remain at rest) and the symmetrical body moves in the medium.

The resultant downward acceleration is independent of the following factors, of course the densities $D_m$ and $D_b$ remain constant.

(i) **The mass of body:** The resultant downward acceleration depends only upon the density of the body and not on mass at all as in Eq. (7.20). It does not depend upon the mass of the body, in fact the resultant downward acceleration is obtained from the equation of the resultant weight by dividing it with the mass of the body in both sides of Eq. (7.11). Thus, the mass disappears from both sides. Consequently, the distance travelled is also independent of the mass and hence the shape. It only depends on the density of the body. The bodies of aluminum of mass 10 kg (distorted) and 10 gm (spherical) or 1 mg (flat) should fall through 27.765 m in 3 s in water but this prediction has not been experimentally confirmed, even at the macroscopic level.

(ii) **The shape of body:** The resultant acceleration as in Eq. (7.20) and hence the distance travelled depend upon the density of the body only, it does not depend on the shape of the body. Also, the distance travelled is independent of the mass of the body. It implies a body of mass of 1000 kg flat in shape and a body of mass spherical in shape must fall equal distances in equal intervals of time. All bodies will travel the same distances in the same time. These equations and concepts became feasible after 1935 years, in this period the principle became established qualitatively in explaining rising, falling and floating bodies.

(iii) **Non-accelerated motion**: The equations of the distance travelled are derived for the case in which the motion is uniformly accelerated. The equation for the distance travelled is based upon the equation

$$S = ut + \frac{1}{2}at^2 = \frac{1}{2}at^2 \tag{7.15}$$

which is applicable if the acceleration is constant. The body falls down with constant acceleration $g$

Now replacing $g$ by $g_R$

$$S = ut + \frac{1}{2}(1 - \frac{D_m}{D_b})gt^2 = \frac{1}{2}(1 - \frac{D_m}{D_b})gt^2 \tag{7.24}$$

All the equations are meant for uniform acceleration. They are not meant for non-accelerated motion.

(iv) **Size of the medium:** Let a fluid fill a tank of water having a width 2 m and a height of 6 m. Now the motion of bodies can be checked in two halves of 3 m each or in three of two meters each. In each half the upward or downward distances travelled by the body must be the same. Likewise, the height of the tank may be taken equal to 50 cm and width to 15 cm to notice the effect of the magnitude of media on the motion of bodies. The magnitude of the medium may be reduced depending upon the size of the body. The body rises upward due to the upthrust against gravity. In the case of rising bodies the magnitude of the medium will be significant. A wooden body of mass 10 gm, can be observed in a test tube, a beaker or a drum or tank of water.

(v) **Viscosity and fluidity:** As already mentioned, according to the Archimedes principle, the viscosity of the medium does not play any role. The coefficient of viscosity of glycerine is 1058 times than that of water. Further, the density of mercury is 13.6 times more than water. In the Archimedes principle, the fluidity of water and mercury are regarded as the same. But this aspect can be checked experimentally. If the distances (upward and downward) travelled are not the same then it may be due to fluidity (normally it means tendency to flow).

**Peculiar comments**

It is now evident that the distance travelled (upward or downward) is the multiple of $g$, i.e. 9.8 m/s$^2$ as in Eq. (7.24). The bodies fall down with the acceleration due to gravity precisely equal to $g$ as a result of gravitational attraction. It is experimentally observed that bodies fall down with the acceleration equal to $g$ in vacuum. However it is not observed in specific experiments that the bodies rise upward with constant acceleration in media. According to the equation based on Archimedes, when the body rises in any medium, e.g. water (1000 kg/m$^3$), glycerine (1260 kg/m$^3$) and mercury (13 600 kg/m$^3$), the distance is the multiple of 9.8. Is it due to the reason that when the body rises upward the gravitational force decreases? Anyhow, the best way to address the problem is to understand it experimentally.

## 4.0. The Archimedes principle and floating bodies

A body floats in water or medium if it neither falls down nor rises upwards. When a body is completely immersed in a fluid then the following forces act on the body.

(i) The weight of the body acts downward

$$W = VD_b g \tag{7.9}$$

(ii) The upthrust experienced by the body in the fluid acts in the upward direction

$$U = VD_m g \tag{7.10}$$

Both forces (weight and up thrust) compete with each other and in some cases they may be equal. If both weight and upthrust are equal then

$$\text{weight} = \text{upthrust}$$
$$VD_b g = VD_m g$$
$$D_b = D_m \tag{7.14}$$

Density of body = Density of medium

Under this condition, both the resultant downward acceleration $g_R$ and resultant downward weight $W_R$ are zero and the body is regarded at rest. When the acceleration of the body is zero, either body is at rest or moves with uniform velocity. Here the second possibility is excluded and the first possibility is favoured. According to it when the body floats (completely submerged) its shape is completely insignificant by all means. There is no term in the equation of the floating body which may account for the shape of the body, the viscosity of the medium etc. The density of the body and that of the medium are the only two terms the appear in the equations.

### *4.1. First stage observations: Floating balloons in water*

Consider an air-filled balloon with a suitable sheath and a tank of water. The balloon is allowed to move upward from the bottom of the tank of water. The balloon rises upward due to the buoyant force or upthrust. If additional weights (masses) are attached to the balloon, then its upward speed decreases as the weight increases. If the weights are judiciously attached then a stage comes in which the balloon floats. Now two types of masses are encountered in the understanding of floating of the balloon in water.

(a) The mass which the balloon experimentally supports. It can be done by adding all the masses attached to the balloon.

(b) The mass can be calculated mathematically noticing various parameters theoretically.

The description is presented in details in Chapter 4.

Let us briefly describe floating balloons of the umbrella and long pipe shape. In the first stage experiments [2] the mass which the balloon supports is found to depend upon the shape of the balloon. For example, an umbrella-shaped balloon (1000 cm$^3$) supported more mass than the predictions of the Archimedes principle. It means if the balloon is umbrella-shaped then more mass was attached externally to stabilize it in the tank of water. Thus, the mass was slightly more than predictions based upon Archimedes. It further implies that the body may float depending upon the shape of the balloon. Secondly if the volume of the balloon is long pipe-shaped (1000 cm$^3$ say) then it supported less mass compared to the case when the balloon was umbrella-shaped. The volume of the balloons of both the shapes is the same (2000 cm$^3$).

Thus, to account for the shape of the balloon, the upthrust is regarded as proportional to the weight of the fluid displaced. Thus the principle has been generalized [2].

$$U \propto \text{weight of fluid displaced by body}$$

The proportionality can be removed by introducing the coefficient of proportionality.

or $$U = fD_m Vg \tag{7.7}$$

where $D_m$ is the density of the medium, $V$ is the volume of the fluid displaced and $g$ is the acceleration due to gravity. All important factor $f$ is the coefficient of proportionality. It depends on the experimental conditions.

The coefficient of proportionality to be pictured is regarded as accounting for the shape of the balloon and other relevant factors. To understand critically the completely submerged floating bodies, the bodies of a non-hygroscopic plastic [8] should be fabricated. The bodies of various shapes and densities slightly less or more than that of water should be fabricated to conduct experiments. Such experiments should be conducted in fluids of high density (mercury) and viscosity (glycerine). It has to be seen whether the fluidity of mercury (13 600kg/$m^3$) is critically same as that of water (1000 kg/$m^3$) or not.

## *4.2. Factors not taken in account by the Archimedes principle*

From the mathematical equations it is evident that the Archimedes principle does not take into account the significant factors regarding the body and the medium.

**Body:** The principle does not take in account the shape and mass of the body, as these two factors do not appear in the mathematical equations. In the mathematical equations based on the Archimedes principle for the rising, falling and floating bodies only the density of body and that of the medium are significant. The mass and shape of the body do not affect the results at all.

These two factors do not occur in any equation at all as there are no factors in equations which may account for the mass and shape of the body.

**Medium**: The principle does not take into account the magnitude of the medium, the motion of the medium, convectional currents, the surface tension and viscosity of the medium, etc. An important liquid in this regard is mercury. A body of cork must rise in water through 144.6 m in 3 s and 2354.94 m in mercury. This prediction can be checked with reduced time $t$ and taking a medium of a different size. Here the fluidity of mercury is taken equal to that of water. From this observation we can conclude whether in the case of fluids the density is only the factor which may affect the results. Also, the fluidity of the medium may be a significant factor. Similarly, observations can be taken in the most viscous or unusual fluids.

**Airplane:** According to the Archimedes principle, the body floats if

$$\text{Density of body} = \text{Density of medium}$$

or $$D_b = D_m \tag{7.14}$$

A steel needle falls in water being heavier than air, i.e. density of the needle is more than that of air. Hence, the condition of flotation based upon the Archimedes principle is obeyed.

A helicopter or airplane floats in air. The density of the helicopter or air plane is more than that of air. So it is direct violation of the Archimedes principle. Why is it not noticed? There are many other factors which are not taken into account by the principle. Such additional factors should have been in mathematical equations. Also it is pertinent to mention here that the airplanes were discovered centuries after the Archimedes principle.

***Anomalous observations***

(i) **The most sensitive experiments:** Similar observation is made in very sensitive experiments. The Archimedes principle is contradicted (as denser particles float over the lighter medium) in sedimentation and ultracentrifugation experiments as reported by R. Piazza [4]. A striking example of this is the rather weird effect that gold particles (spherical in shape) with a density of 19 200 kg/m$^3$ are seen to float atop the lighter surface. The density of gold does not vary in the medium. It is justified that the Archimedes principle is the only limiting approximation and generalization. The Archimedes principle has been generalized by adding the term the effective excess buoyancy to the upthrust.

According to the Archimedes principle, the body floats if

$$\text{Density of body } (D_b) = \text{Density of medium } (D_m)$$

As a gold particle (19 200 kg/m$^3$) is found to float in a lighter medium (say water), then

$$19\,200 \text{ kg/m}^3 = 1000 \text{ kg/m}^3 \text{ (incorrect)}$$

which is not justified. Thus, the Archimedes principle is contradicted in this case. Similar experiments were reported by Heras [5] and published in the Nature Scientific Reports. So these observations are repeatable and experiments are in progress.

## 5.0. Stokes law

This law was proposed by the Irish theoretical physicist and mathematician Sir George Gabriel Stokes (1819–1903). Stokes spent his entire career at the University of Cambridge, where he served as the Lucasian Professor of Mathematics from 1849 until his death in 1903. Stokes made seminal contributions to fluid dynamics. He was the secretary and then president of the Royal Society.

At that time it was high time for Stokes to find out equations for the resultant weight as in Eq. (7.3) and the resultant acceleration as Eq. (7.20). Thus, the distance travelled could have been calculated as in Eq. (7.24). But Stokes confined his studies to a narrow range under some postulates and different phenomena. The famous Stokes law relates viscous force $F$ with various factors and is given by

$$F = 6\pi\eta r v \tag{7.51}$$

where $\eta$ is the coefficient of viscosity, $r$ is the radius of the small sphere, and $v$ is the constant velocity with which the body falls. This basic equation is used in studies in fluid dynamics. The above equation was derived in 1845 and is applicable under five assumptions.

(i) When the body is a small sphere
(ii) Falls through a medium of infinite extent
(iii) The medium is viscous
(iv) The medium in which the body falls remains at rest
(v) The body falls with slow velocity

### *5.1. Constant velocity attained by the falling body*

According to the Stokes law, when the body falls through a viscous medium, it experiences the viscous force. In addition to this, the body experiences a downward force due to weight and upthrust in the medium in the upward direction due to the Archimedes principle. The magnitude of the viscous force increases with increase in velocity. A stage comes in which the viscous force balances the net force (no force acts on the body), then the body starts moving with a constant velocity called terminal velocity $v$. The weight of the small sphere, the upthrust experienced by the sphere due to the medium and the viscous drag act in the following way.

(i) The weight of the body acting vertically downwards which is given by Eq. (7.3) as

$$W = mg = D_b V g \tag{7.3}$$

The weight of the body acts in the downward direction, as attracted by the Earth.

(ii) The upthrust experienced by the body in the medium of density $D_m$

$U$ = The weight of the fluid displaced by the body

$$= D_m V g \tag{7.10}$$

The upthrust acts in the upward direction when the body is immersed in the fluid.

(iii) The viscous drag force acts in the direction opposite to the one in which the body moves. As the body moves downwards the viscous force acts in the upward direction.

$$F = 6\pi\eta r v. \tag{7.51}$$

The viscous force acts in the upward direction as the body falls downwards.

**The three forces compete**

As the body falls down its downward speed keeps on increasing due to gravity as the motion is accelerated. The force of attraction increases as the body approaches the Earth. Consequently, the opposite viscous drag ($F \propto v$), which acts in the upward direction also increases, thus the variation in velocity decreases. Further up the thrust acts upward. Then a stage is reached when the weight of the body is just equal to the sum of the upward thrust and the upward viscous force. At this stage there is no net force to accelerate the body. Thus the body may be at rest or moving with constant velocity. Here it is assumed that the body moves with constant velocity. Hence the body starts moving with constant velocity. When the body attains constant velocity then

$$U + F = W \tag{7.52}$$

Upthrust + Viscous force = Weight
Total upward force = Total downward force

$$D_m V g + 6\pi\eta r v = D_b V g \tag{7.53}$$

In the Stokes law, the small spherical body is considered (not bodies of different shapes) thus its volume $V$ is

$$V = \frac{4}{3}\pi r^3 \tag{7.54}$$

Now substituting the value of volume from Eq. (7.54) into Eq. (7.53) we get

$$D_m \frac{4}{3}\pi r^3 g + 6\pi\eta r c = D_b \frac{4}{3}\pi r^3 g$$

$$6\pi\eta r v = D_b \frac{4}{3}\pi r^3 g - D_m \frac{4}{3}\pi r^3 g$$

or $$6\pi\eta rv = (D_b - D_m)\,\frac{4}{3}\pi r^3 g \qquad (7.55)$$

$$v = \frac{2r^2\left(D_b - D_m\right)g}{9\eta} \qquad (7.56)$$

Equation (7.56) can also be used in understanding the motion of bodies in various fluids. But it is applicable in very-2 narrow range of parameters or variables.

For example

(i) If the density of the falling body is same as that of the medium in which the body falls, i.e.

$$D_m = D_b \qquad (7.14)$$

Now Eq. (7.56) becomes

$$v = \frac{2r^2\left(D_b - D_b\right)g}{9\eta} = 0$$

(ii) If $D_m > D_b$ then the body rises in that medium (velocity is negative, indicated upward motion)

(iii) If $D_m < D_b$ then the body falls in the medium (velocity is positive)

Thus, the Stokes law gives qualitative explanation to rising, falling and floating bodies just like the Archimedes principle, but is applicable under five postulates. The distance travelled by the small sphere is given by Eq. (7.14)

$$S = vt$$

or $$S = vt = \frac{2r^2\left(D_b - D_m\right)g}{9\eta}t \qquad (7.57)$$

## *5.2. Subtle range of Stokes law*

In the Stokes law the additional force is the viscous force, $F = 6\pi\eta rv$. It arises in the case of viscous fluids only. Arnold in 1910 confirmed that the Stokes law is valid for particles of rose metal with a radius of 0.002 cm, i.e. $V = 33.524 \times 10^{-9}$ cc in water with an accuracy of a few tenths of one percent. Thus, it is applicable in an extremely narrow range. A sphere with a radius greater than 0.002 cm falls with a variable velocity or motion. It is also evident from the fact that the Stokes law is valid under five postulates. If further experiments (involving higher masses and different liquids) are conducted in this regard then the range of validity of the Stokes law can be further understood. Further Stokes has not as yet been confirmed in the case of rising bodies. These experiments are significant in the sense that the Stokes law is regarded as the basic law in fluid mechanics and has been extended a number of times. But such experiments will lead to clear understanding of the Stokes law.

## 6.0. Drag force

The total resistance of a plate (body) in the direction of a fluid stream is called the drag force and its magnitude is given by,

$$D = CD_m Au^2 \tag{7.58}$$

where $C$ is the drag coefficient, $A$ is the area of cross-section of the body, and $u$ is the relative velocity (constant and high in magnitude) of the body in the medium. Thus, both the body and the medium are in motion due to external force. Equation (7.58) is applicable when velocity $v$ is *uniform velocity.* The drag force is calculated for moving submarines, airships, cars, cyclists, etc. where external forces are involved. Their motion is not natural motion. The magnitude of $u$ is very high and external forces contribute. Thus, the drag force is not applicable to the natural motion of bodies (bodies fall under the effect of weight in the medium). When bodies fall down in media their speeds are variable (acceleration is constant) as in Eq. (7.20). Thus, $u$ can be positive or negative or zero. In the case of natural motion of bodies the drag force is not applicable.

***Some authors apply drag force to natural motion***

As mentioned above, the drag force acts under the influence of external force. But Resnick [9] applied the same for natural motion also. The drag force depends upon the area of cross-section $A$, and $u$ is relative velocity (constant and high in magnitude) of the body in the medium. For the sake of completeness it is being discussed below. Resnick has equated the weight with the drag force

$$\text{Weight} = \text{Drag force}$$

$$VD_b g = CD_m Au^2 \tag{7.59}$$

Now the constant velocity becomes

$$u = \left(\frac{VgD_b}{CAD_m}\right)^{1/2} \tag{7.60}$$

If $D_m = D_b$, i.e. when the body floats, in this case the velocity is not zero:

$$u = \left(\frac{Vg}{CA}\right)^{1/2} \tag{7.61}$$

When the density of the body is equal to that of the medium ($D_m = D_b$), then the body floats. But according to Eq. (7.61) the body moves with non-zero velocity. Conceptually, how far it is correct is a different issue. But Resnick has neglected a significant factor– upthrust – if the same is taken into account then we get

weight = upthrust + drag force

The drag force is under application of external force.

We have already written a similar equation for viscous force.

weight = upthrust + viscous force

In this case no external force acts on body.

In some sense Resnick regarded the drag force and the viscous force equivalent. However, the ranges of applicability and characteristics of both forces are entirely different.

Thus, the drag force has similar nature like that of the viscous force.

$$VD_b g = VD_m g + CD_m Au^2 \tag{7.62}$$

$$u = \left(\frac{Vg(D_b - D_m)}{CAD_m}\right)^{1/2}$$

$$u = \left(\frac{VD_b g(1 - \frac{D_m}{D_b})}{CAD_m}\right)^{1/2} \tag{7.62}$$

This equation is not applicable in vacuum where all bodies fall down with the same acceleration. Thus, setting $D_m = 0$

$$u \text{ (uniform velocity)} = \frac{VD_b g}{0} = \infty$$

Therefore, the equations are not consistent. In vacuum the body falls with the uniform acceleration or variable velocity.

## 7.0. Developments after the Stokes law

The study of motion of bodies in fluids continued after the Stokes law was proposed and mainly focused on the extension of the conditions or assumptions which underpin the Stokes law along with other effects. Thus, the mathematical equations were extended. This explanation is given for the sake of completeness of discussion of the existing theories. However, this discussion is not relevant to the natural motion of rising and falling bodies.

The equations of motion for small rigid spheres (of particular shape) in the non-uniform flow are discussed in details in the existing literature. Thus, the studies mainly focus on falling spheres and not rising bodies. Originally, Basset [10] and later Boussinesq [11] and Oseen [12] examined the motion of a sphere settling under gravity in a fluid at rest. The disturbance of the flow produced by the motion of the sphere was assumed to be at sufficiently low Reynolds numbers so that the fluid force could be calculated from the results of the unsteady Stokes law. Tchen [13] extended this work first to a sphere settling under gravity in a fluid with an unsteady but uniform flow and then to an unsteady and non-uniform flow, with a view to applications to the turbulent flow. Tchen's work is regarded as fundamental in this regard. Further, this mathematical basis is valid to only those cases in which the Stokes law holds good so that it is applicable under special or limited conditions only, as the term corresponding to the Stokes law appears in equations. The more recent comprehensive and complete analysis was published by given by Maxey and Riley [14]. They reanalyzed the inconsistencies of Tchen's equation, as pointed out by Currsin and Lumbley. For a small rigid sphere of radius $a$ and mass $m_p$, instantaneously centered at $Y(t)$ and moving with velocity $V(t)$, the equation proposed by Currsin and Lumbley [15] is

$$m_p \frac{dV_i}{dt} = m_F \left( \frac{Du_i}{Dt} - \nu \nabla^2 u_i \right) - \frac{1}{2} m_F \frac{d}{dt} \left( V_i[t] - u_i \langle Y(t), t \rangle \right)$$

$$-6\pi a\mu \left[ V_i(t) - u_i\left(Y\langle t \rangle, t\right) + a \int_{-\infty}^{t} d\tau \frac{d/dt \langle V_i(\tau) - u_i \left| Y(\tau), \tau \right| \rangle}{\left[ \pi\nu(t-\tau) \right]^{1/2}} \right] + (m_p - m_F) g_i \tag{7.63}$$

The undisturbed flow field is $u_i(x,t)$, $m_F$ is the mass of the fluid displaced by the sphere, and dynamic and kinetic viscosity are $\mu$ and $v$ respectively. In the above equation it is important to note the distinction between two time derivatives. For a steady flow a further correction is given by Faxen so that the force on the sphere produced by the disturbance flow around the sphere is

$$F = 6\pi a\mu\left[u_i\left(Y\langle t\rangle,t\right)-V(t)\right]+v\pi a^2\left(\nabla^2 u_i\right)_{Y(t)} \tag{7.64}$$

If the distance fallen by the small sphere has to be calculated from Eq. (7.63), then velocity, $v$ ($S = vt$) and acceleration, a ($S = \frac{1}{2}at^2$) has to be calculated which is quiet a tedious process, hence not applied in the existing literature. But it is not so with the generalized theory which is developed below.

In above conditions the motion of the falling small sphere (not bodies of different shapes) is considered in the non-uniform flow and does not aim to give a specific equation for the 'distance' travelled by the body in 'time' $t$. The Stokes law is involved in the equation, hence the equation is applicable in the limited region. Nothing has been mentioned further about the rising and floating bodies. So this study is unidirectional as it ignores the rising and floating bodies. Now our aim is to find out the distance travelled by rising or falling body in the fluid in time $t$ and, in addition, the conditions of flotation in a single formulation. Initially, for simplicity an attempt is made to understand it for the fluid at rest and then in motion. In the current generalized theory all factors (magnitude and shape of the medium, state of motion, convectional currents, viscosity, surface tension and fluidity, the magnitude of the body, shape, distortion of the body, angle at which the body is dropped) are taken into account which makes it applicable for a wide range with mathematical simplicity. The current discussion represents a more general and complete theory on the rising, falling and floating bodies.

## 8.0. Rigorous requirement of generalized theory

Thus it is amply clear in the existing literature that there is no theory which may explain quantitatively the motion of rising, falling and floating bodies. An attempt has been made to develop a theory (the simplest one) to explain the natural state of rising, falling and floating bodies taking all relevant factors into account. In the existing physics

The Archimedes principle and Stokes law are the two principles mainly used for explaining the motion of rising, falling and floating bodies.

(i) The Stokes law is applicable under five postulates if a small sphere moves with **constant velocity**.

(ii) The Archimedes principle states that bodies fall with **constant acceleration** (but with reduced magnitude compared to *g*) in media.

(iii) The drag force is calculated for moving submarines, airships, cars, cyclists, etc. where the external forces are involved. Practically, in the drag force the external force acts on the body (as it moves with uniform velocity), hence it is not applicable in the natural motion of bodies. It is well known that the Archimedes principle and Stokes law are applicable under specific conditions.

(vi) Originally Basset [10] and later Boussinesq [11] and Oseen [12] examined the motion of a sphere settling out under gravity in a fluid at rest. The disturbance of the flow produced by the motion of the sphere was assumed to be at sufficiently low Reynolds numbers so that the fluid force could be calculated from the results of the unsteady Stokes law. In the existing literature the motion of the rigid sphere moving with velocity $V(t)$ is described in Eq. (7.63) in a complex way. The distance fallen by the body in the particular time is not mentioned. In addition, there is no mention about the rising and floating bodies.

However, the generalized theory addresses the phenomena of the rising, falling and floating bodies in the simplest way, as it is specifically formulated for all purposes. Thus it is advantageous and far simpler than the existing perceptions.

But the observation why a steel body with a mass of 1 mgm falls more slowly in water than a steel body with a mass of 10 kg cannot be explained on the basis of the equations (7.20) or (7.24). The reason is that the equations of the distance travelled are independent of the mass and shape of the body. So spherical bodies with a mass of 1 mgm and 100 kg should fall the same distance as the flat body with a mass of 1 gm. Theoretically the bodies may rise upward with constant acceleration in the media. In addition, the equations must be applicable if bodies fall with constant acceleration or variable acceleration. The conditions of applicability of the Stokes law are different. It is applicable if body falls with constant velocity or zero acceleration. So both laws are applicable in the respective ranges.

Archimedes' principle predicts that the bodies of aluminium of mass 10 gm (distorted) and 10 gm (spherical) or 1 mgm (flat) should fall through 27.765 m in 3 s in water but this prediction has not been experimentally confirmed, even at the macroscopic level. Also, according to the Stokes law, the bodies fall with a constant velocity.

So the motion of bodies is not a uniform process as according to the Archimedes principle the bodies fall with a constant acceleration and the Stokes law implies that the bodies fall with constant velocity. Hence, to explain the phenomena over a wide range (motion may be with constant velocity or constant acceleration or variable acceleration), taking all factors into account (i.e. mass, shape and angle at which the body is dropped, magnitude, characteristics of motion of the medium and convectional currents, etc.), a generalized theory of the rising, falling and floating bodies has been formulated as below and can be further refined by conducting experiments.

## *8.1. Background of a generalized or complete theory*

An alternate or complete theory of the rising, falling and floating bodies has been theoretically formulated taking into account the limitations of the existing theories. Here the approach may initially appear postulate-like. An attempt is made to find out the distance travelled by body in time *t* in natural motion. By natural motion we mean that the body falls under the influence of gravity alone and the other external forces do not act on it. Here an attempt is made to find out the distance travelled in time *t*. Some factors are simultaneously included to explain the phenomena which are not taken into account in Archimedes' principle and the Stokes law. Both are applicable under limited and specific conditions. These elusive factors i.e.

(i) magnitude of the medium,
(ii) shape of the medium,
(iii) state of motion of the medium,
(iv) convectional currents,
(v) viscosity of the medium,
(vi) surface tension of the medium,
(vii) fluidity of the medium,
(vii) magnitude of the body
(viii) shape of the body,
(ix) distortion of the body,
(x) angle at which the body is dropped

are very significant but not accounted for in the existing literature.

The Archimedes principle, Stokes law and drag forces only take into account the density of the body ($D_b$) and the density of the medium ($D_m$), the other factors are not taken into account. It is evident from the previous discussion that the resultant acceleration, the resultant weight and the distance travelled are given by

$$a_R = (1 - \frac{D_m}{D_b})g \tag{7.20}$$

$$W_R = (D_b - D_m)Vg \tag{7.11}$$

Further, the distance travelled by the body is given by

$$S = ut + \frac{1}{2}(1 - \frac{D_m}{D_b})gt^2 \tag{7.24}$$

If a body starts from the rest then $u = 0$ and the first term becomes zero. From the critical analysis of various concerned theories it is clear that scientists have made no serious attempts to develop a complete theory of the rising, falling and floating bodies. In the current literature there are simply guidelines as to when the body rises, falls or floats. These guidelines are purely qualitative, not quantitative. However, our aim is to find out

(a) How much distance is exactly travelled by bodies of aluminium (2700 kg/m$^3$) of any shape and mass in water (1000 kg/m$^3$) and glycerine (1260 kg/m$^3$) in 2 s or time $t$? The distance predicted by the equations must be experimentally confirmed.

(b) How much distance is exactly travelled by bodies of cork (240 kg/m$^3$) of any shape and mass in water (1000 kg/m$^3$) and glycerine in 2 s? The distance predicted by the equations must be experimentally confirmed.

(c) In the existing literature there are no such equations which may describe the situation as above, except those presented in the discussion. Anyhow, let us briefly discuss what the Archimedes principle, Stokes law, drag force and advanced fluid mechanics say about this.

**The Archimedes principle:** This principle was stated for the measurement of densities of bodies. But it not clear when its applications were also extended to the rising, falling and floating bodies. In the case of the rising, falling and floating bodies it simply explains the phenomena qualitatively.

If the density of the body is greater than that of the medium, i.e.

$$D_b > D_m$$

then the body falls.

If the density of the body is less than that of the medium, i.e.

$$D_m > D_b$$

the body rises.

If the density of the medium is equal to that of the body, i.e.

$$D_m = D_b \tag{7.19}$$

then the body floats.

This explanation does not give any hint to answer the question about the distances travelled by the body in time. So based upon the Archimedes principle some equations have been formulated i.e. Eq. (7.20) and Eq. (7.24). The various predictions from these equations are shown in Table I. Apparently, one significant limitation of the equations is that they do not account for the shape of the body. Also the motion of bodies is dependent of the mass which is only true in vacuum and not in media. So some sensitive experiments are required to establish them. Further, the equations based on the Archimedes principle are only applicable when the motion is accelerated, i.e. the velocity of the body increases regularly in regular intervals of time.

**Stokes law:** As the Archimedes principle is valid for accelerated motion (velocity is variable), likewise the Stokes law is valid for uniform motion (acceleration = 0). This transformation from variable velocity to constant velocity is caused by the viscous force. The Stokes law takes into account a subtle effect which arises due to the viscous force. The applicability of the Stokes law has been confirmed by Dr. Arnold for small spheres of rose metal with a radius of 0.002 cm or $V = 33.52 \times 10^{-9}$ cc in water for a few tenths of 1%. So it is applicable in the extremely small region and has not even been fully confirmed in the region. Further, the Stokes law is applicable under five postulates. The equations of constant velocity and the distance travelled is given by

$$S = vt = \frac{2r^2\left(D_b - D_m\right)g}{9\eta}t \tag{7.57}$$

Thus both the Archimedes principle and the Stokes law are applicable in specific ranges and not in general.

**Drag force:** Practically speaking, the drag force is not applicable to the natural motion of bodies. For example, it is applicable to the motion of vehicles and vehicles do not move under the effect of gravity. External agents are required to pull the vehicles. When the external force is applied then it is a dynamic system, whereas Eqs. (7.15)–(7.18) are kinematic in nature.

The total resistance of a plate (body) in the direction of the fluid stream is called the drag force [9] and its magnitude is given by,

$$D = CD_m Au^2 \tag{7.58}$$

where $C$ is the drag coefficient, $A$ is the area of the cross-section of the body, and $U$ is relative velocity (constant and high in magnitude) of the body in the medium. It is evident from the definition that the velocity must be constant and high. If the velocity is constant then acceleration is zero. It is has already been mentioned that the mathematical equations based upon the Archimedes principle, i.e. Eq. (7.20), are not applicable if acceleration is zero ($a = 0$ or $u = v$). Now the distance travelled (when velocity $V$ is constant) by the body under the influence of the external force is given by

$$S = Vt$$

Thus, it does not represent the case of a cork body (2.40 kg/m$^3$) which rises upward in water in 2 s or a gold body (19 200 kg/m$^3$) which falls in mercury(13 600 kg/m$^3$).

**Developments after the Stokes law:** Let us consider the widely accepted formulation in this regard. For a small rigid sphere of radius $a$ and mass $m_p$, instantaneously centred at $Y$ (t) and moving with velocity $V(t)$, the equation proposed by Currsin and Lumbley [15] is

$$m_p \frac{dV_i}{dt} = m_F\left(\frac{Du_i}{Dt} - \nu\nabla^2 u_i\right) - \frac{1}{2} m_F \frac{d}{dt}\left(V_i[t] - u_i\langle Y(t), t\rangle\right)$$

$$-6\pi a\mu\left[V_i(t) - u_i\left(Y\langle t\rangle, t\right) + a\int_{-\infty}^{t} d\tau \frac{d/dt\langle V_i(\tau) - u_i|Y(\tau),\tau|\rangle}{\left[\pi\nu(t-\tau)\right]^{1/2}}\right] + \left(m_p - m_F\right)g_i$$

(7.63)

The undisturbed flow field is $u_i(x,t)$, $m_F$ is the mass of the fluid displaced by the sphere, and dynamic and kinetic viscosity are $\mu$ and $\nu$ respectively. In the equation above it is important to note the distinction between the two time derivatives. For the steady flow further correction is given by Faxen so that the force on the sphere produced by the disturbance flow around the sphere is

$$F = 6\pi a\mu\left[u_i\left(Y\langle t\rangle, t\right) - V(t)\right] + \nu\pi a^2\left(\nabla^2 u_i\right)_{Y(t)} \tag{7.64}$$

If the distances fallen by small sphere has to be calculated from Eq. (7.63), then velocity ($S = vt$) and acceleration ($S = at^2/2$) have to be calculated for general cases (mass may be 0.1 gm or less and more than 1 kg) which is quite a tedious process, hence not applied in the existing literature. But it is not so with the generalized theory which is developed below.

## 9.0. Force exerted by fluid or medium $F_m$

The fluid (liquid or gas) has a natural tendency to flow. The solid body may be in any fluid or medium, primarily it is under the effect of gravity and consequently it is attracted by the Earth. The weight of the body is given by $W = VD_b g$. When the body tends to move downwards, the body exerts force $F_b$ on the medium by virtue of its weight. According to Newton's third law of motion, to every action there is an equal and opposite reaction (the generalized form of the third law of motion implies the magnitude of action and reaction may or may not be the same). So as the body exerts a force on the medium, the medium also exerts a force ($F_m$) on the body. The force exerted by the medium (may be in motion, of any magnitude, viscosity etc.) includes the upthrust ($VD_m g$) and other relevant factors. The magnitude of $F_m$ (which includes upthrust, $VD_m g$ and other factors) may be regarded as proportional to the density of the medium $D_m$ mainly.

$$U = VD_m g \tag{7.10}$$

or $$U \propto D_m$$

$$F_m \propto U$$

hence $$F_m \propto D_m$$

The proportionality can be removed by introducing the coefficient of proportionality as

$$F_m = a_m D_m \tag{7.65}$$

The coefficient $a_m$ is obtained by removing the sign of proportionality. There are many coefficients of proportionality in the existing literature, e.g. like the coefficient of thermal conductivity, the coefficient of viscosity, the coefficient of expansion, the drag coefficient etc. It is established that the magnitude of the coefficient of proportionality depends upon inherent characteristics of the process. Here the value of $a_m$ depends on the characteristics of the medium. For the sake of understanding, the value of $a_m$ can be expressed in one of the simplest way as (it can also be expressed in different suitable ways),

$$a_m = x_m\, y_m \tag{7.66}$$

where $x_m$ accounts for the magnitude and shape of the medium. Further, $y_m$ is its state of motion along with other relevant factors like temperature, viscosity, convectional currents, surface tension, fluidity, etc. Ideally, the medium must be at rest. The density of mercury is 13.6 times more than that of water (liquid) but its fluidity (free movements and allowance of bodies to pass through it) may be different from water. Only sensitive experiments can confirm the results or conclusions either way. The specific and quantitative experiments may be conducted in mercury (metal) and other normal fluids (liquids or gases). If the results vary then the reason would be different characteristics of mercury.

***Justification of $a_m$ (factors related to medium)***

The coefficient $a_m$ accounts for the contributions related to the medium which are not taken into account by the Archimedes principle.

## *9.1. Understanding effect of $x_m$*

The significance of the effect of the magnitude of the medium (i.e. $x_m$) can be understood in the following way. In falling bodies, Stokes' law has been reasonably well studied but in a limited range only [7], in this case the magnitude and state of motion of the medium has been clearly defined (i.e. infinite in extent and at rest). If the motion of spheres has been studied in containers then the effect of the walls of the containers leads to a correction factor. It is confirmed that bodies fall with constant acceleration $g$ in vacuum only due to gravity. Also, the bodies may fall with constant acceleration in fluids (but with a reduced magnitude depending upon values of $D_m$ and $D_b$) under certain conditions as already mentioned.

If this prediction is experimentally confirmed under certain conditions then the magnitude of the medium and the shape of the container will be significant. According to the principle, a cork body (say, $r$ = 1 cm) should rise through 3.8701 m in 0.5 s. Let there be four tanks filled with water having dimensions 10 m × 10 m × 10 m, 5 m × 5 m × 5 m, 4 m × 4 m × 3.8701 m, and 0.4 m × 0.4 m × 3.8701 m. Now it has to be experimentally confirmed whether in all cases the cork body rises upward through 3.8701 m in 0.5 s or not as predicted by the Archimedes' principle. If the prediction is not confirmed in some cases then it would be due to the magnitude of the medium only. Similar observations may be made with bodies of different shapes of wood, wax, plastic in water, glycerine and mercury etc.

In the case of falling bodies of different shapes the effect of the magnitude of the medium can be studied at different depths. The Archimedes principle is only valid if the medium is at rest. But the generalized theory is applicable even if the medium is in motion and this effect is taken into account in $y_{m.}$ The incorporation of this factor requires careful experimentations and study.

## 9.2. Understanding effect of $y_m$

As already mentioned, $y_m$ takes into account its state of motion along with other relevant factors like temperature, viscosity, convectional currents, surface tension, fluidity, etc. Due to viscous force ($F = 6\pi\eta rv$) the bodies attain constant velocity under the feasible conditions. At 20°C the density of water is 998.23 kg/m$^3$ and that of glycerine nearly 1260 kg/m$^3$. The coefficient of viscosity of water at 20°C is 0.00101 deca poise and that glycerine 1.069 deca poise. Thus, the coefficient of viscosity is 1058.4 times more than that of water and the density is only 1.2622 times more. As small spheres fall with constant velocity in viscous media (not with constant acceleration) the effect of the viscosity of the medium is significant and justified.

It was confirmed by Arnold in 1910 while justifying the Stokes law [16] that a small sphere of rose metal with a radius of 0.002 cm falls in water with constant velocity with the accuracy of a few tenths of one percent.

(i) Now it can be checked which sphere and under which conditions fully justifies the Stokes law experimentally.

(ii) If the radius of the sphere is less than 0.002 cm whether it falls with constant velocity or not.

(iii) If the radius of the sphere is more than 0.002 cm whether it falls with constant velocity or not.

Till date the Stokes law $F = 6\pi\eta rv$ has been often used in fluid dynamics like $F = ma$ or $E = mc^2$ but it has not been basically confirmed.

(iv) Then it can be confirmed whether the law is obeyed or not. As it is applicable to all viscous media, so different experiments are required to be conducted in all media.

The surface tension is significant as a small thin foil (of metal) rests over the surface of water whereas a small sphere of the same mass sinks in water.

**Table III.** Additional coefficients taken in account in alternate theory

| Sr. No | Coefficient | Significance |
|---|---|---|
| 1 | $x_m$ | Magnitude and shape of medium |
| 2 | $y_m$ | State of motion, convectional currents, viscosity, surface tension and fluidity |
| 3 | $x_b$ | Size of body |
| 4 | $y_b$ | Shape, distortion of body, angle at which body is dropped |

## 10.0. Force exerted by the body ($F_b$) on the medium

When a body is placed freely in any medium, it is attracted by the Earth. It has its weight $W = mg = D_bVg$. Thus, due to the weight the body (irrespective of mass, shape, distortion and angle at which the body is dropped in the medium) the body exerts force $F_b$ on the medium.

The magnitude of force $F_b$ exerted by the body (irrespective of mass, shape, distortion and angle at which the body is dropped in medium) on the medium depends upon the weight

$$W \propto D_b$$

$$F_b \propto W$$

Thus,

$$F_b \propto D_b$$

$$F_b = a_bD_b \tag{7.67}$$

The coefficient $a_b$ is obtained after removing proportionality (like the coefficient of viscosity, the coefficient of thermal conductivity, the coefficient of thermal expansion, etc.), depends on the characteristics of the body and may be expressed in one of the ways as (it can be expressed in different suitable ways also),

$$a_b = x_by_b \tag{7.68}$$

where $x_b$ accounts for the magnitude of the body, $y_b$ for the shape or distortion of the body, the angle at which it is dropped and other relevant factors.

***Justification of $a_b$ (factors related to body)***

The coefficient $a_b$ accounts for the contributions related to the body which are not taken into account by the Archimedes principle and may play a significant role in rising, falling and floating.

## *10.1. Understanding the effect of $x_b$*

**In vacuum**: The value of coefficient $x_b$ depends upon the mass of the body. The effect of the mass of the body can be assessed in two ways i.e. in vacuum and fluids. It has been experimentally confirmed, even in recent sensitive experiments, that bodies fall with the same acceleration $g$ in vacuum only [5]. Thus, in complete vacuum a minute particle of steel of mass of a few macrograms or less and 50 kg or more (irrespective of shape) should fall through equal distances in equal intervals of time (i.e. with the same acceleration). This prediction is based upon Eq. (7.20).

$$a_R = (1 - \frac{D_m}{D_b})g \tag{7.20}$$

Further, the distance travelled is given by

$$S = \frac{1}{2}(1 - \frac{D_m}{D_b})gt^2 \tag{7.24}$$

If bodies (mass of a few macrograms or less and 50 kg or more (irrespective of shape) is dropped in vacuum ($D_m = 0$)

$$a_R = g \tag{7.23}$$

$$S = \frac{1}{2}gt^2$$

The equations (7.20) and (7.24) are independent of mass. So all bodies fall with the same acceleration in vacuum and thus travel equal distances in time $t$.

**In fluids:** In fluids, apart from Eqs. (7.20) and (7.24), another equation is also applicable, i.e

$$v = \frac{2r^2\left(D_b - D_m\right)g}{9\eta} \tag{7.56}$$

Equation (7.56) is applicable under certain conditions in a narrow range. Arnold confirmed it for small spheres (a particular shape) and of particular mass ($33.524 \times 10^{-9}\ D_b$ gm) that fall in water with constant velocity or zero acceleration (the essence of Stokes' law). The accuracy of Arnold's experiments is few tenths of one percent. Such experiments can be conducted with wide range of parameters or variables. A similar equation for a small rising sphere (if the viscous force in the Stokes law is applicable as in the case of falling bodies) can be written as

$$v = \frac{2r^2\left(D_m - D_b\right)g}{9\eta} \tag{7.69}$$

It is equally possible that bodies may fall or rise with variable acceleration depending upon the mass and distortion of the body. Arnold confirmed the Stokes law in the case of falling bodies for spheres with a radius of 0.002 cm. What is the value of the radius of the sphere for which the Stokes law holds good for rising bodies? It is to be determined yet. Such experiments are to be conducted yet. Hence in fluids effects of the mass of the body have to be taken into account. Such experiments can be confirmed in highly dense and viscous fluids. Currently, there are no specific experimental data in this regard.

## *10.2. Understanding the effect of* $y_b$

The value of $y_b$ depends on the shape or distortion of the body, the angle at which it is dropped, and other relevant factors.

To understand the effect of the shape ($y_b$) in a concrete way, consider two steel bodies (7800 kg/m$^3$) having masses of 62.4 gm each, i.e. $V$ = 8 cc. Let one body is a cube of with each side being 2 cm, i.e. 2 cm × 2 cm × 2 cm and the other body is a thin foil. i.e. 50 cm × 16 cm × 0.01 cm. Now this flat body falls slowly in water contrary to the Archimedes principle, i.e.

$$S = \frac{1}{2}(1 - \frac{D_m}{D_b})gt^2 \tag{7.24}$$

According to Eq. (7.24), if the density of bodies is the same then all bodies must fall the same distances in the same time in liquids. This equation does not account for shape (in this case the cube and thin foil) of the body and the magnitude of the body (64.2 gm or 64.2 kg or 1 mgm). Thus it is the limitation of Eq. (7.24).

This effect is purposely accounted for by $a_b$ in an alternate or complete theory. Let us practically analyze the situation when a thin foil (50 cm × 16 cm × 0.01 cm) and cube (2 cm × 2 cm × 2 cm) are placed on water in a tank.

(i) In case of thin foil (50 cm × 16 cm × 0.01 cm) its weight (i.e. the measure of force exerted by the body), and hence $F_b$ on the unit area of water is considerably lower. The reason is that in the case of thin foil its mass is distributed over a large surface area. In case of the cube the surface area is smaller.

(ii) In the case of the cube (2 cm × 2 cm × 2 cm) its weight (i.e., measure of force exerted by the body) and hence $F_b$ on the unit area of water is more considerable compared to the flat thin foil.

But below the unit area of the body (thin foil and cube) the water column (say, a tank of water) remains the same. The force $F_m$ exerted by the medium (water) on the effective unit area of the body (a thin foil and cube) remains the same. So in case of thin foil $F_b$ (force exerted by the body i.e. thin foil) per unit area decreases but $F_m$ (force exerted by the medium) remains the same. Hence the thin foil compared to the cube of same $D_b$ and $V$ falls considerably more slowly in water. The force exerted by the cube per unit area is greater compared to the thin foil. Thus, the force exerted by the medium ($F_m$) appears to be more on the thin foil than that on the cube. To draw concrete conclusions about rising bodies some specific and quantitative experiments are required, which have not been conducted yet.

### *10.3 Units and dimensions of $a_b$ and $a_m$*

We have Eq. (7.65) which relates force with density

$$F_m = a_m D_m \tag{7.65}$$

$$a_m = F_m / D_m$$

or $$a_m = N/\text{kgm}^{-3} = \text{kgms}^{-2}\,\text{m}^3/\text{kg} = \text{m}^4\text{s}^{-2}$$

The units of $a_b$ and $a_m$ are $N$ kg$^{-1}$ m$^3$ or m$^4$ s$^{-2}$ in the SI system and cm$^4$ s$^{-2}$ in the CGS system. The dimensions of $a_b$ and $a_m$ are $M^0L^4T^{-2}$. The values of $x_m$, $y_m$, $x_b$ and $y_b$ are required to be different for different liquids. The exact values can be only determined experimentally.

## 11.0. Comparative study of $F_m$ and $F_b$ or hidden ratio

The rising, falling and floating bodies are determined by the fact that the body exerts a force on the medium and the medium exerts a force on the body. If the force exerted by the body is more than that exerted by the medium then the body falls. If the force exerted by the medium is more than that exerted by the body then the body rises in that medium. The ratio of magnitudes of $F_m$ and $F_b$ is called the hidden ratio (HR). Thus, divide Eq. (7.65) by Eq. (7.67) we get,

$$\text{HR} = \frac{F_m}{F_b} = \frac{a_m D_m}{a_b D_b} = \frac{x_m y_m D_m}{x_b y_b D_b} \tag{7.70}$$

## *11.1. Explanation for motion of bodies*

The hidden ratio is nothing but the ratio of $F_m$ (force exerted by the medium on the body) and $F_b$ (force exerted by the body on the medium). Both forces compete with each other. The state of motion is explained on the basis hidden ratio.

(i) If the hidden ratio is more than one, then the force exerted by the medium dominates the force exerted by the body; hence the body rises in that medium. For example a cork body (240 kg/m$^3$) rises in water (1000 kg/m$^3$).

(ii) If the hidden ratio is less than one then the force exerted by the body dominates that exerted by the medium and the body falls in that medium. For example an aluminium body (2700 kg/m$^3$) falls in water(1000 kg/m$^3$).

(iii) If the hidden ratio is equal to unity then forces exerted by the body and the medium on each other are equal and the body floats. For example, a balloon floats in the atmosphere.

The falling factor (FF) and the rising factor (RF) are measures of the tendency of the body to fall or rise. Thus higher the FF and RF, higher is the tendency of the body to fall or rise. For simplicity the FF and RF can be calculated as

$$\text{FF} = 1\text{–HR} = (1 - \frac{x_m y_m D_m}{x_b y_b D_b}) \tag{7.71}$$

$$\text{RF} = \text{HR–1} = (\frac{x_m y_m D_m}{x_b y_b D_b} - 1) \tag{7.72}$$

The rising factor and falling factor can be calculated by subtracting one from the hidden ratio or the hidden ratio from one. If the hidden ratio is unity, then the body neither falls nor rises.

**Standard conditions.** Under certain conditions (may be termed as standard conditions) the values of $a_b$ and $a_m$ are regarded as unity for simplicity. Thus, under the standard conditions HR, FF and RF become

$$\text{HR} = \frac{D_m}{D_b} \tag{7.73}$$

$$FF = (1 - \frac{D_m}{D_b}) \tag{7.74}$$

$$RF = (\frac{D_m}{D_b} - 1) \tag{7.75}$$

The standard conditions depend on the experimental data and can be understood in the following way.

(i) According to the Archimedes principle, the equation for the resultant acceleration is

$$a_R = (1 - \frac{D_m}{D_b})g \tag{7.20}$$

Thus if $D_m$ and $D_b$ remain the constant then all bodies fall with constant acceleration (variable velocity). The densities can be easily kept constant so all bodies fall with constant acceleration. The conditions under which the bodies fall with constant acceleration may be called standard conditions.

(ii) Also, some contradictory results from the Archimedes principle are obtained when the effect of viscosity is taken into account. According to the existing literature, bodies fall with constant velocity as in Eq. (7.52), i.e.

$$v = \frac{2r^2\left(D_b - D_m\right)g}{9\eta} \tag{7.56}$$

The body falls with constant velocity under five postulates. Thus, according to the existing physics these may be regarded as standard conditions. These postulates are regarded as standard conditions, i.e. if these postulates are satisfied then values of $a_m$ and $a_b$ may be regarded as unity. The conditions under which these predictions are obeyed need to be determined.

(iii) This prediction may be true under certain conditions. Also, in some cases the bodies may fall with variable acceleration.

There is no equation which describes the above situation. However, these predictions are yet to be experimentally confirmed precisely regarding the extent to which the the equations are obeyed. This prediction may be true under certain conditions. Under these three types of motion the standard conditions are different. The conditions, under which these predictions are obeyed, need to be determined.

## 12.0. Standard conditions on the basis of alternate theory

The standard conditions of the body and media are discussed in the forthcoming discussion. These are conditions of the body and media (mass, magnitude etc.) under which the equations are likely to be obeyed. For the specification of standard conditions sensitive experiments with the most sophisticated equipment must be conducted. These conditions may be relative.

(i) **Standard conditions for falling bodies:** If the bodies fall with constant acceleration then in one of the ways the body may be regarded as standard if it is of steel (7800 kg/m$^3$) of a mass of 10 gm ($x_b$ = 1 m$^2$ s$^{-1}$) and spherical in shape ($y_b$ = 1 m$^2$ s$^{-1}$). Similarly, a medium may be regarded as standard in one of the ways if water is filled in the cube with each side 5 m or 10 m in size ($x_m$ = 1 m$^2$ s$^{-1}$) and water is at rest ($y_m$ = 1 m$^2$ s$^{-1}$). It may be the simplest way to understand the standard conditions. The different configurations for the standard body and the medium may be considered. If the motion of the bodies is observed in air in a closed hall with each side 10 m long, then the medium may be regarded as standard ($a_m$ = 1 m$^4$ s$^{-2}$). Also, the standard conditions may be attributed in different ways.

(ii) **Standard conditions for rising bodies:** A wood body (600 kg/m$^3$) of a mass of 10 gm ($x_b$ = 1 m$^2$ s$^{-1}$) and spherical in shape ($y_b$ = 1 m$^2$ s$^{-1}$) may be regarded as standard ($a_b$ = 1 m$^4$ s$^{-2}$). Similarly, a medium may be regarded as standard if water at 20°C is poured into a cube of with each side 5 m or 10 m long ($x_m$ = 1 m$^2$ s$^{-1}$) and is at rest ($y_m$ = 1 m$^2$ s$^{-1}$). Thus, the standard conditions are similar or identical in the case of falling bodies. Such conditions may be changed depending upon experimental suitability.

A typical piece of paper does not fall with constant acceleration in air, but a heavy metallic ball of suitable size falls with constant acceleration. However, both bodies fall with constant acceleration in vacuum where no upthrust is experienced. Similarly a metallic flat or distorted piece does not fall with constant acceleration in water; however a heavy metallic body may fall with constant acceleration in air. Thus the coefficient of proportionality, i.e. $x_m$, $y_m$, $x_b$ and $y_b$ has a significant effect and role in the motion of bodies in various fluids.

## *12.1. The displacement in terms of the falling factor and the rising factor*

(i) **Falling bodies:** In an alternate theory, the falling tendency of the body is not expressed in terms of the resultant downward acceleration but in terms of the falling factor. The higher the falling factor, the higher is the displacement ($S$) through which the body falls,

$$S \propto (\text{FF})$$

The body covers a longer distance if it travels for a longer time

$$S \propto (\text{FF})t$$

Combining the above two proportionalities

$$S \propto (\text{FF})t \qquad \text{or} \quad S \propto (1-\frac{F_m}{F_b})t$$

The proportionality sign can be removed by introducing the coefficient of proportionality.

$$S = A\ (1-\frac{F_m}{F_b})t \tag{7.76}$$

$F_m$: Force exerted by the medium on the body (any mass and shape)

$F_b$: Force exerted by the body on the medium (different magnitudes and viscosities, etc.)

Substituting value of $F_m$ and $F_b$ from Eq. (7.65) and Eq. (7.67) in Eq. (7.76)

$$S = A(1-\frac{x_m y_m D_m}{x_b y_b D_b})t, \tag{7.77}$$

where $A$ is the coefficient of proportionality. Its value depends upon the involved experimental variables and has units m s$^{-1}$. The distance travelled by the body in the medium depends upon the density of the body, the density of the medium, the mass and shape of the body along with the magnitude, viscosity and movement of the media. Thus, the falling factor and the distance through which the body falls depends upon the forces exerted by the body and the medium on each other. It further depends on the various characteristics of the body and the medium. If the values of $x_{m,}$ $y_{m,}$, $x_b$ and $y_b$ are regarded as unity or the net effect is unity, then Eq. (7.77) becomes

$$S = A(1-\frac{D_m}{D_b})t \tag{7.78}$$

The bodies may fall with constant velocity, constant acceleration (variable velocity) and variable acceleration; the general Eqs. (7.77) remain the same, only the values of $A$, $x_{m,}$ $y_{m,}$ $x_b$ and $y_b$ vary depending upon the situation. These can be determined experimentally. In the existing literature such experiments have not been mentioned at all. So it can be safely concluded that it is an original problem to be solved and will add to the contents and debates in science.

(ii) **Rising bodies:** In an alternate theory the rising tendency is not expressed in terms of the resultant upward acceleration but in terms of the rising factor. The higher the rising factor in the medium, the higher is the displacement through which the body rises,

$$S \propto (\text{RF})$$

The body covers longer distance if it travels for longer time.

$$S \propto (\text{RF})t$$

Now combining above proportionalities, we get

$$S \propto (\text{RF})t \quad \text{or} \quad S \propto (\frac{F_m}{F_b}-1)t$$

$F_m$: Force exerted by the medium on the body (any mass and shape)

$F_b$: Force exerted by the body on the medium (different magnitudes, densities and viscosities, etc.)

$$S \propto (\frac{x_m y_m D_m}{x_b y_b D_b}-1)t,$$

The proportionality can be removed by introducing a coefficient $B$ as

$$S = B(\frac{x_m y_m D_m}{x_b y_b D_b} - 1)t \tag{7.79}$$

where $B$ is the coefficient of proportionality. Its value depends upon the involved experimental variables and has units m s$^{-1}$. Thus the rising factor and the distance through which the body falls depend upon the forces exerted by the body and the medium on each other. It further depends on the various characteristics of the body and the medium. If the values of $x_{m,}$ $y_{m,,}$ $x_b$ and $y_b$ can be regarded as unity or their net effect is unity, then Eq. (7.79) becomes

$$S = B(\frac{F_m}{F_B} - 1)t = B(\frac{D_m}{D_b} - 1)t \tag{7.80}$$

where $B$ is a coefficient and determined experimentally. Its nature and magnitude are precisely identical to $A$. In the existing literature there is no quantitative data about rising bodies to draw concrete conclusions. So some quantitative experiments are required to ascertain the characteristics of $B$. The bodies may fall with constant velocity, constant acceleration (variable velocity) and variable acceleration; the general Eqs. (7.79) remain the same, but only the values of $B$, $x_{m,}$ $y_{m}$, $x_b$ and $y_b$ vary depending upon the situation. These can be determined experimentally.

### *12.2. Comparison of equations giving distances (upward and downward)*

The distance which the body travels according to the Archimedes principle is

$$S = \frac{1}{2}(1 - \frac{D_m}{D_b})gt^2 \tag{7.24}$$

Also the downward and upward distances in the alternate theory are given by

$$S = A(1 - \frac{x_m y_m D_m}{x_b y_b D_b})t, \tag{7.77}$$

$$S = B(\frac{x_m y_m D_m}{x_b y_b D_b} - 1)t \tag{7.79}$$

According to the Archimedes principle, the distances travelled (upward and downward) are multiples of $g$ (9.8 m/s$^2$), i.e. dependent on $g$. The bodies fall with the precisely constant acceleration $g$, in vacuum and with reduced acceleration in fluids. Also, the body is attracted towards the Earth. However, the situation of the rising bodies is different. In this case bodies are pushed upward in the media, e.g. water (1000 kg/m$^3$), glycerine (1260 kg/m$^3$) and mercury (13 600 kg/m$^3$) and the distance travelled is the multiple of $g$. As the distance is the multiple of $g$ this puts severe constraints on the rising bodies. The distance travelled by the rising body is also the multiple of $g$, like the distance travelled by the falling bodies. When the body rises upward the force of attraction decreases due to the inverse square law of gravity.

However, in the equations based upon the alternate theory the rising, falling and floating bodies are independent of $g$ (9.8 m/s$^2$).

## *12.3. The floating bodies*

The falling factor and the distance travelled by the body when it falls are given by

$$\text{FF} = 1\text{–HR} = \left(1 - \frac{x_m y_m D_m}{x_b y_b D_b}\right) \tag{7.71}$$

$$S = A\left(1 - \frac{x_m y_m D_m}{x_b y_b D_b}\right)t, \tag{7.77}$$

Similarly, the rising factor and the distance through which the bodies rise are given by

$$\text{RF} = \text{HR–}1 = \left(\frac{x_m y_m D_m}{x_b y_b D_b} - 1\right) \tag{7.72}$$

$$S = B\left(\frac{x_m y_m D_m}{x_b y_b D_b} - 1\right)t \tag{7.79}$$

Consider the case when the values of $x_{m,}$ $y_{m,}$ $x_b$ and $y_b$ are unity or their net effect is unity and the density of the body is equal to that of the medium, i.e. $D_m = D_b$. In this case, the values of FF, RF and the distances travelled (downward and upwards) are given by

$$\text{FF} = 1\text{–HR} = (1–1) = 0 \tag{7.81}$$

$$S = A(1–1)t = 0 \tag{7.82}$$

Thus, the body has no falling tendency. The hidden ratio is unity which means that the force exerted by the body on the medium and the force exerted by the medium on the body are equal. Similar results also follow from Eq. (7.72).

$$\text{RF} = \text{HR}-1 = (1-1) = 0 \tag{7.83}$$

$$S = B(1-1) = 0 \tag{7.84}$$

Thus, the body has no rising tendency. The hidden ratio is unity which means that the force exerted by the body on the medium and the force exerted by the medium on the body are equal. Thus, when $x_{m,}$ $y_{m,}$ $x_b$ and $y_b$ are unity (or the net effect is unity), and the density of the medium ($D_m$) is equal to that of the body ($D_b$) and both are equal, then the body floats in that medium.

Whereas Eq. (7.20) implies that when $D_m = D_b$ then the resultant acceleration is zero. When the acceleration is zero, then either body is at rest or moves with uniform velocity. Here scientists regard the preferred case when the body is at rest.

(a) **When bodies fall with constant velocity: the distance travelled by the body in the medium is given by Eq. (7.77).**

$$S = A\left(1-\frac{x_m y_m D_m}{x_b y_b D_b}\right)t \tag{7.77}$$

The velocity can be written from Eq. (7.77) as

$$V_{av} = A\left(1-\frac{x_m y_m D_m}{x_b y_b D_b}\right) \tag{7.85}$$

Equation (7.85) gives the value of average velocity. If the body moves with constant velocity then the average velocity and the constant velocity are the same, i.e. $V = v_c$. This situation occurs in the Stokes law:

$$V_{av} = v_c = A\left(1-\frac{x_m y_m D_m}{x_b y_b D_b}\right) \tag{7.86}$$

Now consider the case when $x_i$'s and $y_i$'s are unity or their net effect is unity. Then Eq. (7.86) becomes

$$V_{av} = v_c = A\left(1-\frac{D_m}{D_b}\right) \tag{7.87}$$

Also we have the equation for constant velocity from Stokes law as

$$v = \frac{2r^2\left(1-\frac{D_m}{D_b}\right)D_b g}{9\eta} \tag{7.56}$$

Thus comparing Eq. (7.87) with Eq. (7.56), as velocity remains constant, we get

$$A = \frac{2r^2 D_b g}{9\eta} \tag{7.88}$$

Now substituting the value of $A$ from Eq. (7.88) in Eq. (7.87)

$$V_{av} = \frac{2r^2\left(1 - \frac{D_m}{D_b}\right) g D_b}{9\eta} \tag{7.89}$$

Thus, we get the equation for the terminal velocity in the Stokes law. So, Arnold verified Eq.(7.56) for small spheres of rose metal with a radius of 0.002 cm in water [16]. The value of $\eta$ for water at 20°C is 0.01 poise then

$$A = 0.8888888 \times 10^{-5} D_b g \tag{7.90}$$

and

$$V_{av} = 0.8888888\ 10^{-5} D_b(1 - \frac{D_m}{D_b})g \tag{7.91}$$

From Eq. (7.91) the velocities for steel and aluminium bodies in water at 20°C are 0.05923 cm/s and 0.014808 cm/s. Thus, there is a complete agreement between predictions of Eq. (7.91) based upon the alternate theory and the existing experimental findings.

**(b) When bodies fall with constant acceleration:** It has been confirmed that the bodies fall with constant acceleration equal to $g$ in vacuum [7]. Also, according to Archimedes' principle, i.e. Eq. (7.20), the bodies fall with constant acceleration but with reduced magnitude in fluids depending upon the values of $D_m$ and $D_b$ (in this case shape and mass are no factors which can influence results). So far it has not been experimentally confirmed; yet under certain conditions it may be true. Both the cases are discussed as below.

**(i) When bodies fall with constant acceleration in vacuum.** In this case the displacement depends upon time as $t^2$ as in Eq.(7.24), i.e.

$$S = ut + \frac{1}{2}(1 - \frac{D_m}{D_b})gt^2 \tag{7.20}$$

$$S = \frac{1}{2}(1 - \frac{D_m}{D_b})gt^2 \tag{7.24}$$

Also the distance through which body falls is given

by $$S = A\left(1-\frac{x_m y_m D_m}{x_b y_b D_b}\right)t, \tag{7.77}$$

It is observed that when the body falls with constant acceleration then the distance depends upon the square of time, i.e. $t^2$. Thus, we can write

$$A = kt \tag{7.92}$$

where the value of $k$ depends upon the experimental conditions. Or

$$S = k\left(1-\frac{x_m y_m D_m}{x_b y_b D_b}\right)t^2 \tag{7.93}$$

If the values of $x_{m,}$ $y_{m,}$ $x_b$ and $y_b$ are regarded as unity (or the net effect is unity), then Eq. (7.93) becomes

$$S = k\left(1-\frac{D_m}{D_b}\right)t^2 \tag{7.94}$$

Comparing Eq. (7.94) with Eq. (7.24)

$$k\left(1-\frac{D_m}{D_b}\right)t^2 = \frac{1}{2}\left(1-\frac{D_m}{D_b}\right)gt^2 \tag{7.95}$$

$$k = \frac{1}{2}g \tag{7.96}$$

Putting Eq. (7.96) in Eq. (7.94)

or $$S = \frac{1}{2}\ g\left(1-\frac{D_m}{D_b}\right)t^2$$

which is nothing but Eq. (7.24). It is obtained from the generalized theory. Thus, if we regard the values of $x_{m,}$ $y_{m,,}$ $x_b$ and $y_b$ as unity (or the net effect) we get the same result as the one obtained using the Archimedes principle.

***In vacuum***

In vacuum, the density of the medium can be regarded as zero, i.e. $D_m = 0$, then Eq. (7.77) becomes

$$S = At \tag{7.97}$$

It is observed that when the body falls with constant acceleration then the distance depend upon the square of time, i.e. $t^2$. Thus we can write

$$A = k_0 t \tag{7.98}$$

$$S = At = k_0 t^2 \tag{7.99}$$

where the value of $k_0$ depends upon the experimental conditions.

Also, in the existing physics the distance travelled is given by the equation

$$S = \frac{1}{2}(1 - \frac{D_m}{D_b})gt^2$$

If the body falls in vacuum, then $D_m = 0$

$$S = k_0 t^2$$

Now comparing both equations of the distance travelled in vacuum.

$$k_0 t^2 = \frac{1}{2} g t^2$$

$$k_0 = \frac{1}{2} g$$

or

$$S = \frac{1}{2} g t^2 \tag{7.100}$$

which is the equation in existing physics. Thus, from the alternate theory of the rising, falling and floating bodies we get the consistent results which are found in the existing physics. Here it is assumed that all the coefficients, i.e. $x_{m,}$ $y_{m,}$ $x_b$ and $y_b$ are unity or their net effect is unity.

Thus, in the case of falling bodies the results of the alternate theory and from the equations based upon the Archimedes principle are the same. The only difference is that in the alternate theory the effects of the magnitude and shape of the medium, its state of motion along with other relevant factors like temperature, viscosity, convectional currents, surface tension, fluidity, etc., are taken into account. In addition, the alternate theory takes into account the magnitude of the body, the shape or distortion of the body, the angle at which it is dropped, and other relevant factors. This is the main difference between the existing equations and the alternate theory.

## 13.0. The falling bodies and generalized theory

According to the generalized theory, the bodies fall down if the hidden ratio is less than one, i.e. the force exerted by the body is more than that exerted by the medium. The measure of the tendency of the body to fall is the falling factor, given by Eqs. (7.71) and (7.74); thus the higher the FF, the higher the tendency of the body to fall.

(i) **In vacuum.** In complete vacuum $D_m$ can be regarded as zero and also HR is zero; this is less than one hence the bodies fall down. Further, the falling factor for all bodies in vacuum is unity, hence they possess the equal and maximum tendency to fall. Thus, according to Eqs. (7.77) all bodies travel equal distances in equal intervals of time in vacuum. In the language of the current science all bodies fall with the same acceleration. It is justified from Eq. (7.100) which is originated from Eq. (7.77).

(ii) **In air.** The HRs from Eq. (7.25), i.e. under standard conditions:

$$\text{hidden ratio} = \frac{D_m}{D_b} \tag{7.73}$$

$$\text{falling factor} = (1 - \frac{D_m}{D_b}) \tag{7.74}$$

(d) For a body of aluminum (2700 kg/m$^3$) in air (1.293kg/m$^3$)
hidden ratio = 0.0004788 (7.101)
falling factor = 0.999251 (7.102)

(b) For a body of steel (7800 kg/m$^3$) in air (1.293 kg/m$^3$)
hidden ratio = 0.0001657 (7.103)
falling factor = 0.999834 (7.104)

(c) For a body of silver(10500 kg/m$^3$) in air (1.293 kg/m$^3$)
hidden ratio = 0.0001231 (7.105)
falling factor = 0.999877 (7.106)

(d) For a body of platinum (21 500 kg/m$^3$) in air (1.293 kg/m$^3$)
HR = 0.00006013 (7.107)
falling factor = 0.9999398 (7.108)

As the values of HRs are less than unity, i.e. force exerted by the body dominates, thus the bodies fall in the medium. Also, the falling factors are less than one and non-zero. But our ultimate aim is to calculate the distance travelled by the body in time *t*.

## *13.1. Standard conditions*

Under the standard conditions ($x_{m,}$ $y_{m,}$ $x_b$ and $y_b$ or the net effect is unity) Eq. (7.77) becomes

$$S = A(1-\frac{D_m}{D_b})$$

$$t = k_m(1-\frac{D_m}{D_b})t^2 \quad (7.109)$$

(i) A steel body in air

Let it be that a steel body of mass 10 gm ($x_b = 1$ m² s⁻¹) and spherical in shape ($y_b = 1$ m² s⁻¹) is regarded as standard and it falls through a distance of 100 cm in time $t$. Let $k_m = k_{sa}$ for the steel body air. Then Eq. (7.109) for the steel body air becomes

$$S = 100 = k_{sa}\ (1-\frac{D_{air}}{D_{steel}})\ t^2 = k_{sa}\ (0.999834)t^2$$

or $$k_{sa} = 100/0.999843t^2 = \frac{100}{0.999843t^2} \quad (7.110)$$

An aluminium body in air can be regarded as standard (in one way) in air as below. Let the value of $k_m$ for the aluminium body in air is $k_{aa}$ and is equal to $k_{sa}$ (steel in air)

$$\text{i.e.}\quad k_{sa} = k_{aa}$$

if falls with constant acceleration in air. Let it be that in air the aluminium body falls through the distance $S$, then Eq. (7.109) with the help of Eq. (7.110) becomes ($k_{sa} = k_{aa}$)

$$S = k_{sa}\ (1-\frac{D_{air}}{D_{alu}})\ t^2 = k_{sa}\ (0.999521)t^2$$

$$= \frac{100}{0.999843t^2}\ (0.999521)t^2 = 99.9677 \quad (7.111)$$

Thus the aluminium body spherical in shape will be regarded as standard in air which travels aluminium distance equal to 99.9677 cm in time $t$ in which the steel body of mass 10 gm, spherical in shape, falls through 100 cm. It is just possible that this prediction may be confirmed for the aluminium body of the mass precisely equal to 10 gm or nearly equal to 10 gm (the shape of the body can be slightly different from spherical).

(ii) **In water:** Just like air, the standard conditions can be determined whether the body will rise in water or not.

**Table IV.** Hidden ratios and falling factors for various bodies in various fluids under standard conditions are shown in Table IV

| Sr. No | Body | FF = 1–HR in air | FF = 1–HR in water |
|---|---|---|---|
| 1 | Aluminium (2700 kg/m$^3$) | 0.999251 | 0.629629 |
| 2 | Steel (7800 kg/m$^3$) | 0.999834 | 0.8717949 |
| 3 | Silver (10 500 kg/m$^3$) | 0.999877 | 0.9047619 |
| 4 | Platinum (21 500 kg/m$^3$) | 0.999939 | 0.953488 |

**Table V.** Hidden ratios and falling factors for various bodies in various fluids under standard conditions are shown in Table V

| Sr. No | Body | RF = HR–1 in air | RF = HR–1 in water |
|---|---|---|---|
| 1 | Wood elm (600 kg/m$^3$) | 0.997845 | 0.66666 |
| 2 | Cork (250 kg/m$^3$) | 0.994828 | 3 |
| 3 | Wax (900 kg/m$^3$) | 0.998563 | 0.111 |
| 4 | Willow wood (420 kg/m$^3$) | 0.996921 | 1.381 |

It can be determined by calculating the hidden ratio and falling factors in water.

(a) For an aluminium body (2700 kg/m$^3$) in water (1000 kg/m$^3$)

hidden ratio = 0.3703703 (7.112)

falling factor = 0.629629 (7.113)

(b) For a steel body (7800 kg/m$^3$) in water

hidden ratio = 0.128205 (7.114)

falling factor = 0.871794 (7.115)

(c) For a silver body (10 500 kg/m$^3$) in water

hidden ratio = 0.095238 (7.116)

falling factor = 0.9047619 (7.117)

For the bodies of aluminium, steel, silver and platinum in water the hidden ratios are less than one hence the bodies falls down. As the falling factors for the bodies are less in water than in air so the bodies fall more slowly in water. But it is required to be quantitatively confirmed by relating them to the distance travelled. It means the exact distance travelled by the body in time must be measured. But when comparing the motion of the same body in two different fluids the fluidity (tendency to flow or allowance of bodies to pass through it) must be taken into account.

**(ii) Steel in water**

As in the previous case, the bodies can be regarded as standard. The aluminium body can be regarded as standard in water in the following way. For a standard steel body ($a_b = 1$ m$^4$ s$^{-2}$), in water ($a_m = 1$ m$^4$ s$^{-2}$, i.e. the tank of water with each side equal to 5 m and water is at rest) if it travels the distance of 100 cm in time $t$, the Eq. (7.109) becomes ($k_{aw} = k_{sw}$)

$$100 = k_{sw}\,(1 - \frac{D_w}{D_{steel}})\,t^2 = k_{sw}\,(0.8717948)t^2$$

or

$$k_{sw} = \frac{100}{0.871794t^2} \tag{7.118}$$

Here $k_{sw}$ is the value of $k_m$ for the steel body in water. Let the value of $k_m$ for the aluminium body in water be $k_{aw}$ which is equal to $k_{sw}$, i.e.

$$k_{aw} = k_{sw},$$

if the aluminium body falls with constant acceleration. If in the tank of water with each side 5 m long ($x_m = 1$ m$^2$ s$^{-1}$) and water is at rest ($y_m = 1$ m$^2$ s$^{-1}$), the aluminium body falls through the water the distance $S$ in time $t$. Then the distance $S$ from Eq. (7.109) becomes ($k_{aw} = k_{sw}$)

$$S = k_{sw}\,(0.630285)t^2 = \frac{100}{0.871794t^2}\,(0.630285)t^2 = 72.278 \text{ cm} \tag{7.119}$$

Now that the aluminium body in water can be regarded as standard and falls through the distance 72.28 cm in time $t$, in which the steel body of mass 10 gm spherical in shape falls through 100 cm in the same time. If this prediction is nearly confirmed for the aluminium body of mass 10 gm, then slight deviations from the spherical shape can be considered for the precise confirmation. The distance travelled in time can be quantitatively measured by measuring the value of $k_m$. It requires some specific and sensitive experiments involving various bodies and fluids. Thus, theoretical discussion may not be conclusive in this regard.

## 14.0. The rising bodies and alternate theory

(i) **In air.** Just like the falling bodies, the state of the rising bodies can also be understood in view of the alternate theory of the rising bodies. In such cases the hidden ratios and rising factors are to be calculated. Consider the motion of hydrogen- and helium-filled balloons in air. For simplicity,

the densities of the balloons are taken the same as those of hydrogen (0.0899 kg/m$^3$) and helium (0.1785 kg/m$^3$).
For hydrogen-filled balloons (0.0899 kg/m$^3$, say) in air (1.293 kg/m$^3$)
hidden ratio = 14.3826 (7.120)
rising factor = 13.3826 (7.121)

For the helium-filled balloon (0.1785 kg/m$^3$, say) in air
hidden ratio = 7.2436 (7.122)
rising factor = 6.2436 (7.123)

As the hidden ratio is more than one the balloons rise upward in water. As the rising factor for the hydrogen-filled balloon is more than that for the helium-filled balloon, hence the former rises upward quickly

**(ii) Justification of Maxwell's distribution law:**
According to Maxwell's law of distribution of molecular speeds [10] lighter gases escape more easily from the Earth's atmosphere. It can be justified on the basis of the generalized theory. Here let us discuss the motion of hydrogen, helium, oxygen (1.428 kg/m$^3$) and carbon dioxide (1.977 kg/m$^3$), considering or assuming each as an individual body in air.

(d) hidden ratio for oxygen (1.428 kg/m$^3$) in air (1.293 kg/m$^3$)
hidden ratio = 0.905462
falling factor = 0.094537

(e) hidden ratio for carbon dioxide (1.977 kg/m$^3$) in air (1.293 kg/m$^3$)
hidden ratio = 0.65402
falling factor = 0.34598

Hence if oxygen and $CO_2$ are considered as bodies, then they fall down in air. Thus, lighter gases ($H_2$ and $He_2$) escape easily from the Earth's atmosphere due to higher $RF_s$.
rising factor for hydrogen: HR–1 = 14.382–1 = 13.382
rising factor for hydrogen: HR–1 = 7.24–1 = 6.24

The heavier gases ($CO_2$ and $O_2$) fall down or settle at the bottom in air due to higher FFs

falling factor for carbon dioxide = 1–HR = 1–0.65402 = 0.34598
falling factor for oxygen = 1–HR = 1–0.905462 = 0.094537

The FFs for these are 0.094537 and 0.34598; hence they fall down in air. Thus, lighter gases ($H_2$ and $He_2$) escape easily from the Earth's atmosphere due to higher RFs, i.e. 13.382 and 6.24; and heavier gases ($CO_2$ and $O_2$) fall down or settle at the bottom due to higher FFs, i.e. 0.34598 and 0.094537. Thus, deductions from the generalized theory are consistent with Maxwell's law of distribution of molecular speeds and also with experimental observations.

(iii) **In water:** Similarly, on the basis of the generalized theory the upwards motion of the cork body (240 kg/m$^3$) and wood (600 kg/m$^3$) can be explained.

For the cork body of (240 kg/m$^3$) in water (1000 kg/m$^3$)
hidden ratio = 4.1666
falling factor = 3.1666

For body of wood (600 kg/m$^3$) in water (1000 kg/m$^3$)
hidden ratio = 1.6666
falling factor = 0.6666

The RFs for bodies of cork (240 kg/m$^3$) and wood (600 kg/m$^3$) from Eqs. (7.75) i.e. under standard conditions, are 3.1593 and 0.66372, hence they rise upward. As the RF for cork is more than that of wood the cork rises upward more quickly than wood. Plastic bodies are fabricated such that they do not absorb water or non-hygroscopic bodies [8] having the densities equal to wood and cork.

**Wood in water**

Let a standard body of wood ($a_b$=1 m$^4$ s$^{-2}$) in water (tank with each side 5 m long and water is at rest, i.e. $a_m$= 1 m$^4$ s$^{-2}$) travels the distance of 100 cm in time $t$. If the value of $B$ for wood in water is $B_{ww}$ then Eq. (7.80) becomes

$$S = B_{ww}\left(\frac{D_{water}}{D_{wood}} - 1\right)t = B_{ww}(0.6666)t$$

or

$$B_{ww} = \frac{100}{0.6666t} \tag{7.124}$$

Then the cork body in water can be regarded as standard in the following way. Let the value of $B$ for cork in water is $B_{cw}$ which is equal to $B_{ww}$, i.e. $B_{cw} = B_{ww}$ and the body rises with constant acceleration. If in the tank of water ($a_m$= 1 m$^4$ s$^{-2}$) a body of cork ($a_b$ = 1 m$^4$ s$^{-2}$) rises through the distance $S$ in time $t$, then the value of $S$ for the cork body from Eq. (7.80) with the help of Eq. (7.24) can be calculated as ($B_{cw} = B_{ww}$)

$$S = B_{ww} \left(\frac{D_{water}}{D_{cork}} - 1\right) t = B_{ww} \ (3.1593)\ t$$

$$= \frac{100}{0.6666t} (3.1593)\ t = 473.942 \text{ cm} \qquad (7.125)$$

Now that the cork body in water ($a_m = 1$ m$^4$ s$^{-2}$) will be regarded as standard which rises through 473.942 cm in time $t$, in which the wood body rises through 100 cm. Now slight variations in the shape of the cork body of mass 10 gm can be considered: so this prediction is precisely confirmed for the body of mass 10 gm.

If this prediction is justified for some other cork body of mass other than 10 gm in a tank of water of the dimensions other than 5 m (water is at rest) then that body and the tank is regarded as standard.

**Possible effect of viscosity on rising bodies:** such experiments can be conducted in highly viscous fluids like glycerine to understand the effect of viscosity in the case of rising bodies. It is experimentally confirmed that due to the viscous force small spheres fall down with constant velocity. It is experimentally confirmed that due to the viscous force the small spheres fall down with constant velocity in fluids. Is a similar effect also observable in the case of rising bodies also? Or the bodies rise upwards with accelerated motion, i.e. up to which extent is the Archimedes principle obeyed? It can be experimentally checked. There are no experimental observations in the existing physics regarding this. The distance travelled in certain time can be quantitatively measured by measuring the value of $B$. It requires some specific and sensitive experiments involving various bodies and fluids. For the rising bodies the magnitude of the medium may be significant as for the same density the bodies may have different masses. A body typical in size and shape may be allowed to rise in different pots. A body with a mass of 1 gm can rise in a beaker having 50 cc of water and in a tank having 1000 litres or more of water. The experiments can be performed in viscous and non-viscous fluids.

## 15.0. The floating bodies and the generalized theory

According to this theory the body floats if HR is equal to unity the and the magnitudes of forces, i.e. $F_m$ and $F_b$, become equal. Thus, in Eqs. (7.71) and (7.72) both FF and RF become equal to zero, hence also the distances travelled according to Eqs. (7.77) and (7.78) and (7.79) and (7.80) are zero.

(i) **In air:** Let a hydrogen-filled balloon has RF equal to 3 (under standard conditions), thus it rises upward. As it rises upward the density of air decreases, hence HR. At some stage the forces exerted by the balloon and the medium on each other become equal so HR becomes unity. Both RF and FF and the distances travelled according to Eqs.(7.73), (7.74), (7.77), (778), (7.79) and (7.80) become zero. Thus the body floats at that stage, i.e. neither rises nor falls.

(ii) A cork body falls in air (FF = 0.994612) and rises upward in water (RF = 3.1593) and rests over the surface of water partially submerged. Small steel pins sink in water due to definite FF equal to 0.872022. But if these pins are pricked to the cork they start resting on the surface of water and the system (cork and steel) is partially submerged. As the pins are continuously pricked to cork the density of the system gradually increases and HR decreases. As soon as the force exerted by the system becomes more than that exerted by water due to gradual pricking of the pins to the cork HR becomes less than one and the body sinks.

## 16.0. Conclusions

In the existing literature the motion of rising, falling and floating bodies are explained by the Archimedes principle, the Stokes law and drag force, but each is applicable in specific and limited ranges. In the Archimedes principle only the density of the body ($D_b$) and that of the medium ($D_m$) are taken into account. In Stokes law $D_m$ and $D_b$ are significant factors, however particles are small in size and spherical. These two laws, i.e. Stokes law and the Archimedes principle, act in the natural motion of bodies (no external or mechanical forces act on the body, only the acceleration due to gravity, upthrust, viscous drag act on body). The drag force ($D = CD_mAv^2$) is applicable when external forces act on the body, e.g. motion of a submarine in water or a car in air, (both run by external fuel), etc., so it is different from natural motion.

**Table VI.** Comparison of existing theories on rising, falling and floating bodies with a generalized theory in natural motion of bodies

| | Characteristics | Existing Theories | Alternate Theory |
|---|---|---|---|
| | | **1. Falling bodies** | |
| (i) | Term | Resultant downward acceleration. $g_R = (1 - \frac{D_m}{D_b})$ | $FF = (1 - \frac{x_m y_m D_m}{x_b y_b D_b})$. |
| (ii) | Displacement | $S = \frac{1}{2} g_R t^2$ ($S = \frac{1}{2} at^2$) | $S = A(FF)\, t$. Independent of $g$ |
| (iii) | Applicable if | $g_R$ is constant. | No such constraint. |
| (iv) | Preliminary predictions | All bodies fall with constant acceleration in fluids if $D_m$ and $D_b$ remain the same. | All bodies may not fall with constant acceleration depending upon $a_m$ and $a_b$. |
| (v) | Typical predictions | All bodies of steel of mass 10 gm (distorted) and 10 gm (sphere) should fall in fluids equal distances in equal times. | These bodies should not fall through equal distances in equal times depending upon $a_i$' s. |
| (vi) | Contradictions | In denser fluids | No such constraint. |
| (vii) | Reasons for contradictions | $g_R$ only depends upon $D_m$ and $D_b$ | Not applicable. |
| (viii) | Stokes' law | Required, if applicable. | Not required. |
| (ix) | Drag force | Not applicable | Not required. |
| | | **2. Rising bodies** | |
| (i) | Term | Resultant upward acceleration $h_R = (\frac{D_m}{D_b} - 1)g$ | Rising Factor $RF = (\frac{x_m y_m D_m}{x_b y_b D_b} - 1)$ |
| (ii) | Displacement | $S = \frac{1}{2} h_R t^2$ | $S = B(RF)t$ |
| (iii) | Applicable if | $h_R$ is constant. | No such constraint. |
| (vi) | Dependence | $h_R$ does not depend on mass and shape of body. | *RF* takes all such factors in account. |
| (v) | Preliminary predictions | All bodies should rise with constant acceleration in fluids if $D_m$ and $D_b$ remain the same. | All bodies may not rise with constant acceleration depending upon $a_m$ and $a_b$. |
| (vi) | Typical predictions | A body of cork of mass 10 gm (distorted) and 10 gm (sphere) should rise in water through 3.8701 *m* in time 0.5 *s*. | These bodies may not rise through 3.8701 *m* in 0.5 *s* depending upon values of $a_m$ and $a_b$. |
| (vii) | Verification | Specific tests have not been conducted so far. | Quantitative tests have been suggested. |

| Characteristics | | Existing Theories | Alternate Theory |
|---|---|---|---|
| (viii) | Stokes' law | Required, if applicable. | Not required |
| (ix) | Drag force | Not applicable | Not required. |
| | | **3. Floating bodies** | |
| (i) | Term | Resultant accelerations, $g_R$ & $h_R$ | RF and FF |
| (ii) | Condition | $g_R$ and $h_R$ are zero | RF and FF are zero |
| (iii) | Equation | $D_m = D_b$ | $D_m = fD_b$ |
| (iv) | Typical prediction | A pallet of mass (flat) of mass 1 gm or less of density 13.60001 gm per cc should sink in mercury. The observation can be repeated with sphere. | The pallet may not sink depending upon values of values of $a_m$ and $a_b$. |
| (v) | Verification | In preliminary floating balloon tests in water it has not been found correct. | This has been justified in these balloons tests. |

In the equation concerning the rising and falling bodies the acceleration due to gravity $g$ (9.8 m/s$^2$) appears in mathematical equations. So the distance travelled (fallen or risen), is the multiple of the acceleration due to gravity, $g$. It is justified in the falling bodies as bodies fall with same acceleration in vacuum (9.8 m/s$^2$), but there is no such observation in the case of rising bodies. Thus multiples of $g$ in distance arisen is justified or not? Thus such experiments are absolutely necessary regarding the rising bodies. The generalized theory is only meant for the natural motion of bodies.

(i) The Archimedes principle states that bodies fall with **constant acceleration** (but with reduced magnitude than $g$) in media. The distance travelled (fallen or risen) according to the Archimedes principle is independent of the shape of the body, the magnitude of the body, distortion of the body and the angle at which the body is dropped. However, in practice these factors are very significant. A steel pin falls in air but an airplane floats in atmosphere; this is not explained on the basis of The Archimedes principle.

(ii) The Stokes law is applicable if a small sphere moves with constant velocity. In practice, this law is applicable for small spheres (radius 0.002 cm, say) in water, i.e. limited and specific range.

(iii) The drag force is calculated for moving submarines, airships, cars, cyclists etc., where external forces are involved (external fuel is used). External force acts in practice in the drag force, hence it is not applicable in the natural motion of bodies. So both existing theories are applicable under specific conditions. Here an attempt is being made to develop a general theory for rising, falling and floating bodies (e.g. 0.1 mgm or less, 50 kg or more; and spherical, flat, thin foil or distorted shapes, may have holes or twists at the surfaces such the that density remains precisely the same) which may be applicable under general conditions. No external forces (fuel engine, etc.) act on the system. Further, elusive factors, i.e. mass, shape and the angle at which the body is dropped, magnitude, characteristics motion of medium and convectional currents etc. must be taken into account. In addition, motion should be explained over a wide range (motion may be with *constant velocity or constant acceleration or variable acceleration*). Thus, the generalized theory has been proposed to explain the motion over a wide range. The existing theories are unable to explain the motion of rising, falling and floating bodies over a wide range.

## References

1. Heath, T.L., *The Works of The Archimedes principle*, C.J. Clay and Sons, pp. 258–261 (1897).
2. Sharma, A., Speculations in Science and Technology, 20, 1997, 297–300.
3. Newton, I., *Mathematical Principles of Natural Philosophy,* pp. 19–20, London, 1727, translated by Andrew Motte from the *Latin*.
4. Piazza R., et al., *Soft Matter*, **8**, 7112–7115 (2012).
5. De las Heras et al., Nature Scientific Reports, 2,789, 2012.
6. Michel, W.C., *Foundations of Physics* (D. Van Nostrand, London), pp. 255–56, (1968).
7. Niebauer, T.M., et. al., Al Phys. Rev. Lett. 59 6 (1987).
8. Bizzeti, P.G., et al., Phys. Rev. Lett. 62, 2901–4, and various references therein (1989).
9. Resnick, R., et al., *Fundamentals of Physics* (Asia Book Pvt. Ltd. New Delhi, 3rd impression,), pp. 138–139 (1995).
10. Basset, A.B., *Treatise on Hydrodynamics* (Deighton Bell, London, 1888), Vol. 2, Chap. 22, pp. 285–297.
11. Boussinesq, J., *Theorie Analytique de la Chaleur* [L Ecole Polytechnique, Paris, 1903], Vol. 2 p.225
12. Oseen, C.W., *Hydrodynamik* (Leipzig, 1927), p.132
13. Tchen, C.M., PhD thesis, Delft, Martinus, Nijhoff, The Hague, 1947
14. Corrsin, S. and Lumbley, J., Appl. Sci. Res., A 6, 114 (1956)
15. Maxey, J.R. and Riley, J.J., Phys., 26(4), April 1983
16. Millikan, R.A, *The Electron* (University of Chicago Press, Chicago), pp. 93–96 (2nd edition, 10th impression).

8

# Route to Newton's Laws of Motion

## First glimpse

- ✓ Aristotle put forth that 'there can be no motion without force' or 'cause and effect perception', but was unable to explain the projectile motion.
- ✓ Philoponus in 6$^{th}$ century and Buridan in 14$^{th}$ century contradicted this assertion and put forth the Impetus Theory.
- ✓ Galileo hypothesized a medium devoid of resistive forces.
- ✓ In early 17$^{th}$ century Galileo preached the law of inertia that the body maintains its uniform motion even without force if once set in motion. Apparently, it was just opposite to Aristotle's assertion and Buridan/Philoponus Impetus Theory.
- ✓ On the other hand, Aristotle, Philoponus and Buridan maintained that the force or impetus (energy imparted) is required for motion; they considered the medium with resistive forces.
- ✓ In 1687 Newton re-stated Galileo's law of inertia and gave the laws of motion in treatise the Principia.
- ✓ Galileo's law of inertia and the Newton first law of motion imply that as long as resistive forces (atmospheric, frictional, and gravitational, etc.) are absent the body moves with uniform velocity and stops when resistive forces start.
- ✓ It would be better if scientists formulate a comprehensive law for practical systems (having resistive forces) and this law must reduce to ideal cases (devoid of resistive forces), such as Galileo's law of inertia.

## 1.0. Aristotle and his dynamics

Aristotle was born in 384 B.C. in Stagira, in Thrace, at the northern end of the Aegean, near Macedonia. Aristotle's father was the family physician of King Philip of Macedonia. He was trained first in medicine and then in 367 he was sent to Athens to study philosophy with Plato. He stayed at Plato's Academy until about 347. Though a brilliant pupil, Aristotle opposed some of Plato's teachings and when Plato died, Aristotle was not appointed head of the Academy.

Aristotle is said to have written 150 philosophical treatises. The 30 that survive touch on an enormous range of philosophical problems, from biology and physics to morals to aesthetics to politics. But here we will be only concerned about the motion of bodies.

He returned to Macedonia in 338 to tutor Alexander the Great. After Alexander conquered Athens, Aristotle returned to Athens and set up a school of his own, known as the Lyceum. After Alexander's death, Athens rebelled against Macedonian rule, and Aristotle's political situation became precarious. To avoid being put to death, he fled to the island of Euboea, where he died soon after.

### *1.1. The theory of cause and effect*

The basic principle of Aristotelian philosophy had been: 'there is no effect without a cause.'

It means no incidence (motion) can occur automatically, there has to be reasons for motion. The object has to be moved by a mover. When the above notion is applied to the motion of bodies, obviously it implies that 'there is no motion without a force.'

Here: effect ≡ motion, cause ≡ force

Another example is that a cart only moves if an ox pulls it. When the ox stops pulling, the cart stops. So the cart moves only as long the ox pulls it by applying the biological force on it. Nowadays petrol-run vehicles are analogous to the ox pulling the cart or a man dragging a table. The vehicle stops moving when the engine is switched off. When the cyclist stops paddling the bicycle stops just after that. The biological energy of the paddler is converted to kinetic energy. The motion of airplanes is again similar.

Thus, the higher the force applied the higher would be the speed of the body.

So Aristotle's ideas were taught over 1800 years and seem true even now, as these looked practical. The reasons are simple, nobody has seen a cart which moves without pull. In the world, no vehicle, no train, no plane etc. have been invented which may move without the external force. Here

**Cause:** External force (as originated in the engine) acting on the vehicle, train, plane etc.

**Effect:** Motion of the vehicle, train and plane

According to Aristotle, when the external force acts on the body, its speed increases.

(i) Higher the external force applied, higher the speed i.e.
Speed $\propto$ force
Higher the force applied more would the speed of the body be. If a lesser force is applied slowly the body moves. Thus, the speed directly depends upon force.

(ii)
$$\text{Speed} \propto \frac{1}{\text{resistance}}$$

The resistance is higher on a rough surface than on a smooth surface. If the body is pushed with the same force on both surfaces, then it moves farther on the smooth surface than on the rough surface. Hence more resistance the smaller the speed or less resistance the higher the speed. Combining above proportionalities

$$\text{Speed} \propto \frac{\text{force}}{\text{resistance}}$$

or
$$\text{force} \propto \text{speed} \times \text{resistance}$$

Aristotle wrote the above proportionality as

$$\text{Force} = (\text{resistance}) \times (\text{speed}) \tag{8.1}$$

Aristotle confined the study to the above equation only. However, the above proportionality can be removed by introducing the coefficient of proportionality as we do now,

$$\text{Force} = K \text{ speed} \times \text{resistance} \tag{8.2}$$

The value of the coefficient of proportionality is determined experimentally; if in all cases it is found to be the same the coefficient of proportionality is called the constant of proportionality. But Aristotle did not investigate the equation quantitatively. *K* may depend upon the distance travelled. Thus, the explanation remained qualitative only. However, Eqs. (8.1)–(8.2) can be completely discussed taking all factors into account, as other equations are discussed.

### *1.2. Motion of a projectile*

An arrow moves in the forward direction in air and it is not directly pulled or pushed by any force, i.e. mover and moved are not in contact. When an arrow flies in air then no external force acts directly on it, i.e. it is not directly in contact with the motion-causing agent. Why does the arrow continue to fly after it has left the bow-string, or a stone after it has ceased to be in contact with the hand that throws it? Since Aristotle supposed that

a) whenever there is motion there must be something which imparts the motion, and
b) the mover and the moved must be in contact, he was led to conclude that the air displaced in front of the projectile somehow rushes round it and pushes from behind, thus propelling the projectile along.

So, according to Aristotle, it is the push of air which causes the arrow to fly in the forward direction even when pushed from the bow. The reason for this assumption is that the central idea of Aristotle's philosophy is that the mover should be in direct contact with the moved.

The cause of motion of the arrow is the push of air. Aristotle's theory of cause and effect is also contradicted as in vacuum an arrow moves without any push or pull. In vacuum there is no air which may push the arrow from behind. But this conclusion, i.e. there has to be the cause of motion, was the basis of the theory of impetus.

## 2.0 Johannas Philoponus's theory of impetus

In the sixth century (490–570), the philosopher Philoponus studied at the school of Alexandria and began publishing scientific articles from about 510. Philoponus became one of the earliest thinkers to reject Aristotle's dynamics, he did not agree with the notion that the medium keeps the projectile (arrow) going. In simple words, an arrow does not move due to the push of air when left from the stretched bow.

Aristotle had used this argument to explain the motion of an arrow in air after leaving the stretched bow.

Philoponus proposed that motion was not maintained by the action of a surrounding medium, but by some property imparted to the object when it was set in motion.

Johannas Philoponus became the first person to reject Aristotle's idea that a body would move only as long it is directly pushed by the mover. He argued that an arrow moves in air when no force acts on it. He proposed the '*Theory of impetus*' i.e. 'an object moves and continues to move because of the *energy imparted* to it by the mover, and ceases movement when that *imparted energy* is exhausted'.

Thus the arrow moves when it leaves the bow, due to the reason that the energy is imparted to the arrow by the bow. When the *energy imparted* to the arrow vanishes the arrow stops. Thus impetus maintains the motion when the external force directly ceases to act, e.g. motion of an arrow.

Philoponus believed that motion is a quality, like the Aristotelian qualities of warmth or moisture. Once a body is set in motion, it will keep moving in the absence of friction because a moving body contains a certain quantity of the 'quality of motion' exhibited by the body due to the *energy imparted* to it. This perception implies that in the absence of resistive forces the body keeps on moving. The frictional forces destroy the energy imparted to the body hence motion. Thus, in the absence of resistive forces the arrow will move continuously. Thus, he perceived the foundations of the concept of inertia before Galileo. Galileo maintained that a moving body stops due to resistive forces, if no resistive forces are present the body keeps on moving. Further, Newton's first law of motion is nothing but the law of inertia. The energy provided to the body previously is preserved in the absence of resistive forces and tends to move with uniform velocity. So Philoponus's doctrine has bearing on Galileo's law of inertia and Newton's first law of motion.

However, Philoponus's theory was largely ignored at the time because he was too radical in his rejection of Aristotle. It may be due to the reason that on the whole people found Aristotle's doctrine more practical. The ideas similar to impetus were expressed by writers such as Hipparchus (c190BC) and Synesius (AD373) but Philoponus never acknowledged that he had knowledge of such ideas.

### *2.1. Papal decree and theory of impetus*

In the 13th century, Thomas Aquinas believed that it is God that keeps our sphere in rotation. Thomas Aquinas took the motion of heavenly bodies as proof for the existence of God. The theory of impetus contradicted the same (the object moves by the energy given by the mover), and was disliked by the holy church. This view was made official in the condemnation of 1277, a papal decree intended to suppress speculation that might contradict the teachings of the church.

Nevertheless, the ideas of Philoponus were adopted by some philosophers in the 14th century. William of Ockham and Jean Buridan called the 'quality of motion' *impetus*. These scholars avoided the problem of the condemnation of 1277 by invoking God as the originator of the impetus, i.e. God is the mover. It was further stated that the almighty God could change the motions if He pleased.

## 3.0. Jean Buridan coined impetus

Jean Buridan (1295–1360), a French philosopher taught at the University of Paris and critically analyzed Aristotle's doctrines. Jean Buridan, who was the chief exponent of 'impetus' dynamics, William of Ockham and Jean Buridan called the 'quality of motion' *impetus*. Although the idea of impetus is credited to Philoponus, the mathematical equation for impetus was given by Buridan as

$$\text{Impetus} = \text{weight} \times \text{velocity} \tag{8.3}$$

According to modern perceptions, the dimensions of the impetus are $ML^2T^{-3}$.

The impetus increases directly with the weight of the body and with its velocity. This concept is similar to Newton's 'quantity of motion' or the modern concept of 'momentum'. At Buridan's time the weight was not defined so mass and weight were the same terms. In view of this Buridan's impetus is just the modern day momentum. Buridan applied the impetus theory to the violent motion of bodies and the falling motion of bodies where Aristotle applied his doctrine. In addition to this Buridan applied the same to celestial motion.

**(a) Celestial motion:** Buridan thought that the celestial bodies, the stars, planets, the sun and moon, were moving in an empty outer space without resistance. Why do they move? What is the cause of motion? Here Buridan applied his impetus theory to explanation of the motion of heavenly bodies. Even at that time people never knew that the Sun is centre of the Earth and planets moves around it. Also at that time scientists, mathematicians and philosophers had no clue about the universal gravitation.

For the motion of heavenly bodies they believed that it is due to impetus. God has given initial impetus to heavenly bodies at the time of creation and bodies will move forever. Thus, celestial bodies move for ever with the original impetus given by God to them at the beginning. Now we say that all the planets move around the Sun due to force of gravitation. Also, we understand that whole universe is expanding.

(b) **Falling bodies or Buridan supported Aristotle's views on falling bodies**

Buridan also used his theory of impetus to explain the behaviour of falling bodies. Unlike many of Galileo's contemporaries, Buridan claimed that stones accelerate as they fall. Buridan assumed that a falling body acquired velocity (impetus) proportional to weight (mass). It implies that the more the mass of body the more quickly it moves. So a 10 pound shot falls more quickly than a 5 pound shot. It is just Aristotle's views on falling bodies, which has been abandoned.

However, the current perception is different as lighter or heavier bodies fall with the same acceleration, i.e. travel equal distances at equal intervals of time. It is an experimental fact that all objects fall at the same speed, on the Earth it is 32 ft. per $s^2$ or 9.8 $ms^{-2}$, regardless of their weight (mass). This aspect is also discussed in chapter 1.

(c) **Violent motion:** Aristotle asserted that the air provided the motive force for the violent motion of a body. Philoponus strongly objected the fact. Looking for an alternative explanation, he concluded that there must be a motive force that resides within the stone itself. When a force was imparted to a stone by another object, the stone absorbed the force, moving a little faster. Thus, the motion of stone, simply continues until air resistance or a collision with another body stopped it. Thus, the stepping stone for Galileo's law of inertia or Newton's first law of motion is laid by Philoponus. Galileo maintained if resistive forces are eliminated, then the body is not stopped. Galileo's law of inertia and Newton's first law of motion are based directly on this doctrine.

Buridan gave this hypothetical force of Philoponus the name 'impetus'. According to Buridan, the 'impetus' depends upon both the speed and the quantity of the mass in a body. He reasoned that even after the body received this impetus force the body would maintain its motion. This motion is maintained after the body loses contact with the mover. In fact, the body being moved, under ideal conditions, would continue to move in a straight line, at uniform speed forever. This idea is recognized as Galileo's law of inertia or Newton's first law of motion.

## 4.0. Galileo's law of inertia

To understand the concept of inertia the preceding scientific developments have to be quickly glanced.

(i) Depending upon daily life observations, Aristotle stated that 'force is required for movement.'
Even now this aspect is supported by experiments. For example a table can be pushed only by applying external force on it. The table does not move automatically, if so it will cause havocs. There are many more such examples.

(ii) But an arrow after leaving the bow moves in air. While it moves in air no direct force acts on it. Aristotle held that air pushes the arrow in air, thus the former is the mover. But this perception was contradicted by many, including Philiponus.
Philiponus assumed that "an object moves and continues to move because of an *energy imparted* to it by the mover, and ceases movement when the *energy imparted* is exhausted." Thus an arrow moves with the *energy imparted* to it by the mover. Further the arrow comes to rest due to air resistance or frictional forces. Then Buridan also continued with the same doctrine.

(iii) Thus an arrow or projectile moves due to the energy imparted to it by the mover, and comes to rest due to frictional or resistive forces. The resistive forces dissipate the energy imparted to the body.
Galileo dealt with these frictional or resistive forces effectively, considering a system devoid of resistive forces. Galileo put forth that if frictional or resistive forces are zero

then the body (an arrow) once set in motion will maintain its uniform motion perpetually.

Thus, Galileo's approach was towards the ideal or hypothetical system or non-existent system (devoid of all resistive forces) on the Earth as the gravitational force can never be negated. In the ideal system the body maintains its state of uniform motion forever. We cannot nullify the resistive forces on the Earth, gravitational force is always present on the Earth.

So it would be better if the laws are formulated for practical systems (resistive forces are present or resistive forces > 0) in general; and under limiting cases (resistive forces = 0) these must reduce to ideal cases. Realistically the prediction must be such that the force or its influence or impact is required for the motion of bodies on the practical system, and in the ideal system the body must maintain its original velocity ($v = u$) forever.

### *4.1. Demonstration of Galileo's law of inertia*

(i) Consider a system of a person having a ball in hand, standing on the rough ground on the surface of the Earth. Let the person throws or pushes the ball on the ground.

(ii) As the person pushes the ball on the rough ground, it travels some distance and obviously stops due to the presence of resistive forces (atmospheric, frictional, gravitational forces etc.). Let the ball travels the distance equal to 5 m on the ground.

(iii) Then the rough pebbles are removed from the ground and extra sand is also cleaned or removed. Then again with the same force the ball is pushed on the ground. In this case it travels a larger distance as resistive forces are somewhat minimized. Now the ball moves a distance of 15 m. Thus as the resistive forces are removed, a larger distance is travelled by the ball.

(iii) Further consider that the surface of the ground is further smoothened, minimizing the frictional and atmospheric forces (still air). Then the ball is thrown with the same force. In this case the ball will travel a longer distance, say 30 m.

As the resistive forces are removed the ball travels a longer distance. It implies that the ball comes to rest due to the presence of resistive forces.

**Experiments on wooden plank**

(iv) Consider smooth wooden planks are laid down on the ground and a smooth ball is considered. If the ball is again pushed with smooth wooden planks are laid down same force then it will travel a much greater distance and maintains uniform velocity for longer time. In this case smooth wooden planks are laid down ball may travel a distance equal to 100 m due to a substantial decrease in resistive forces.

**Experiments on glass floor**

(v) Further consider a more interesting case in which a smooth glass is laid down on the ground. Also a glass ball is taken and pushed with the same force as in the previous cases. Now the ball will maintain uniform velocity up to a long distance and may travel a distance equal to 10 km. In these cases resistive forces are lower than in all cases.

We may minimize the frictional force, atmospheric force but the gravitational force on the Earth cannot be negated. So the ball stops after travelling the distance equal to 10 km on the smooth glass floor.

**Hypothetical system**

(vi) Let us assume that if the gravitational force is also nullified completely along with atmospheric and frictional forces, then the body will maintain its uniform motion perpetually if once set in motion.

This is explanation for Galileo's law of inertia. It can be understood in many ways. The Newton's first law of motion is nothing else but the law of inertia.

**Conclusions**

Hence we find that Aristotle (a force is required by the body to move), Philiponus (the arrow moves due to imparted energy), Galileo (the body maintains uniform motion in the absence of resistive forces) or Newton (the same as in case of Galileo) have studied problems in individual parts. If one aspect is addressed then the other one is neglected.

It would be better if scientists formulate a comprehensive law for a practical system (in the presence of atmospheric force, frictional force, gravitational forces, etc.) and this law must reduce to ideal cases (devoid of resistive forces), such as Galileo's law of inertia. The single law must explain the problems in both systems. The prediction of the mathematical equation should be such that a force

**Table I.** Comparison of views on force and motion of scientists/philosophers Aristotle, Philoponus, Buridan, Galileo and Newton.

| Sr. No | Scientist/ philosopher | Era | Characteristic/ definition | Merit | Limitation/ Region of validity |
|---|---|---|---|---|---|
| 1 | Aristotle | (384-322 BC) | Force is required for motion. Resistive force = 0 | Table moves if pushed by mover | Motion of projectiles, move without contact with mover. |
| 2 | Philoponus | 600 AD | Moves due to energy imparted. Resistive force = 0 | Motion of projectile | No mention of direct contact with system and motion in ideal system |
| 3 | Buridan | 1295–1360 | Same as above | Same as above | Same as above |
| 4 | Galileo | 1564–1642 | Body moves with uniform velocity if resistive force = 0 | Ideal or hypothetical system | Not applicable if resistive forces are present, as motion is not uniform. |
| 5 | Newton | 1642–1727 | Same as above | Same as above | Same as above |
| 6 | Possible Future | onwards | Law must be simultaneously applicable to both resistive and ideal systems. | Should be complete law for resistive and ideal systems. | Applicable for both ideal and practical systems. Should be perfect law |

is required to move the body in the practical system, and the body if once set in motion then must maintain uniform velocity without force in the ideal or hypothetical system. So in the same formulation the explanation for the practical and hypothetical system must be considered. The various theories are shown in Table I.

### *4.2. Linear momentum*

Let us understand what is meant by the concepts momentum and force. These two terms are significant in understanding the dynamics of bodies.

The linear momentum is defined as the total quantity of motion contained in a body and is measured as the product of the mass of the body and its velocity.

The momentum is the product of mass (scalar) and velocity (vector), hence is a vector quantity. It is denoted by $P$. Mathematically,

$$P = mv \tag{8.4}$$

The units of force are:

SI system: kg m $s^{-1}$
CGS system: gm cm $s^{-1}$
Dimensional equation, $P = MLT^{-1}$

(i) In some sense, the concept of momentum was also given by Jean Buridan (1295–1360) as the impetus, i.e.

$$\text{Impetus} = \text{weight} \times \text{velocity} \tag{8.3}$$

At that there was no distinction between weight and mass, thus Eq. (8.3) becomes

$$\text{Impetus} = \text{mass} \times \text{velocity}$$

(ii) Also Newton defined [1] the term the quantity of motion which is also momentum.

'The quantity of motion is the measure of the same, arising from velocity and quantity of matter conjunctly (involving the joint activity of two or more).'

It is obvious that the quantity of motion involves velocity and mass (quantity of matter) collectively. Thus, Newton defined 'quantity of motion' as we understand momentum ($mv$) nowadays.

### *4.3. Force*

The force is the push or pull which either changes or tends to change the state of rest or of uniform motion.

It is our common observation that an object such as chair lying in a room remains at rest unless a push or pull is given to it. Such an object cannot move at its own.

Further a body is moving with some velocity in straight line. It is found that the force is required to change its velocity. Also external force is required to change the direction of the body. In other words, the force is an agent which causes acceleration $a$.

The force is the product of mass (scalar) and velocity (vector), hence is a vector quantity. It is denoted by $F$. Mathematically,

$$F = ma \tag{8.5}$$

The units of force are:

SI system: Newton, kg m $s^{-2}$

CGS system: Dyne, gm cm $s^{-2}$
Dimensional equation, $F = MLT^{-2}$

$F = ma$ is taught globally as Newton's second law of motion. But neither this equation nor such definition was given by Newton in the Principia. It is the biggest irony or paradox in the science even now. There is a clear scientific proof that Newton did not write the equation $F = ma$ in the Principia. We have no proof at all whether Archimedes discovered the principle known after his name in the tub of water and whether Galileo dropped two shots of different masses from the Leaning Tower of Pisa. These anecdotes became prevalent as the scientists got immortal fame. It should be ended with logical conclusions.

Galileo introduced the concept of inertia in 1605, i.e. 82 years before Newton. Galileo Galilei, however, realized that a force is necessary to change the velocity of a body, i.e. acceleration. But no force is needed to maintain its velocity if it is moving with uniform velocity. In other words, Galileo stated that in the *absence* of a force, a moving object will continue moving. The tendency of the object to resist changes in uniform motion or state of rest was what Galileo called *inertia*.

## 5.0. Newton's three laws of motion

Newton's most famous book, popularly known as the Principia, was written originally in Latin as Philosophiæ Naturalis Principia Mathematica in 1687. It was translated to English by Andrew Motte [1] in 1729 as 'Mathematical Principles of Natural Philosophy'. It is a work in three books by Sir Isaac Newton, first published 5 July 1687. The second edition was published in 1713 and third one in 1726. Thus, the third edition of the Principia was published 39 years after the first edition with considerable additions and alternations.

On page 19 of the Principia Newton stated three laws of motion with a short explanation. The most peculiar thing is that the first and third laws, as we teach them now, are exactly the same as stated by Newton, but the second law of motion is different from the law stated by Newton. It means Newton neither stated 'the rate of change of momentum' nor $F = kma$ or $F = ma$. Then who stated it? How did $F = ma$ appear in the Principia? Why are the scientists silent about it? Why no steps are taken to correct it? It will affect the faith of general public that in science everything is consistent. Can the law be gifted to Newton who is the greatest scientist ever walked on the Earth? Then neglecting the facts and remaining silent provides no solution in science. Scientists should come out with facts and credit of the discovery must be given to the genuine inventor. Moreover, Newton's laws in the Principia are terse and telling or speaking, these are not like prophesies of Nostradamus that could be interpreted according to one's convenience.

Two full English translations of Newton's Principia have appeared, both based on Newton's 3rd edition of 1726. Thus, the third edition of the Principia was published 39 years after publication of the first. In every edition there are additions and alternations. In the discussion the first translation as published by Andrew Mott in 1729 is used as it is expected to be closer to Newton's original text. Apparently this translation may have started in Newton's days. This translation is referred in this discussion. Newton's three laws of motion in original Latin and English are given below [1]. Newton's first law of motion is influenced by Galileo's observations of inertia, whereas the second law of motion is the mathematical form of the first law of motion. However, Newton's third law of motion is independent.

## *5.1. Three laws of motion in Latin and English*

***Lex I:*** *Corpus omne perseverare in statu suo quiescendi vel movendi uniformiter in directum, nisi quatenus a viribus impressis cogitur statum illum mutare.*

Translated to English

**Law I:** Every body persists in its state of being at rest or of moving uniformly straight forward, except insofar as it is compelled to change its state by force impressed.

Newton's second law of motion as stated in original Latin

***Lex II:*** *Mutationem motus proportionalem esse vi motrici impressae, et fieri secundum lineam rectam qua vis illa imprimitur.*

Translated to original English

**Law II:** The alteration of motion is ever proportional to the motive force impress'd; and is made in the direction of the right line in which that force is impress'd.

Newton's third law of motion as stated in original Latin

***Lex III:*** *Actioni contrariam semper et œqualem esse reactionem: sive corporum duorum actiones in se mutuo semper esse œquales et in partes contrarias dirigi.*

When translated to English
**Law III:** To every action there is always an equal and opposite reaction: or the forces of two bodies on each other are always equal and are directed in opposite directions.

How Newton explained the laws is discussed in the forthcoming discussion.

## 6.0. First law of motion

**Law I:** Every body persists in its state of being at rest or of moving uniformly straight forward, except insofar as it is compelled to change its state by force impressed.

Now the following inferences can be drawn from the definition of the first law of motion.

(i) If the body is at rest, then it can be set in motion, if force is applied on it. For example a player kicks a football at the rest and football attains velocity.

(ii) If a body is moving with constant speed along straight line, then in order to increase or decrease its speed an external force is required. On the surface of the Earth the speed of the body decreases due to the resistive forces present. Here Newton's law is completely silent about how the body is set to uniform motion. Obviously, an external force is required initially to set the body in motion. Newton should have mentioned the point for completeness.

(iii) If a body is moving with constant speed along straight line, then in order to change the direction of motion, a force has to be applied in the direction normal to the direction of motion.

### *6.1. Explanation by Newton of the first law of motion in the Principia*

Newton has given the following explanation to the first law of motion in the Principia on page 19. It pertains to projectile motion, rectilinear motion and motion of heavenly bodies:

"Projectiles persevere in their motion, so far as they are not retarded by the resistance of the air, or impelled downwards by the force of gravity. A top whose parts by their cohesion are perpetually drawn a side from rectilinear motions, does not cease its rotation, otherwise than as it is retarded by the air. The greater bodies of planets and comets, meeting with less resistance in more free spaces preserve their motions both progressive and circular for much longer time."

## 7.0. Second law of motion

Newton has given the following explanation to the second law of motion in the Principia on page 19. It pertains to forces acted on body under different conditions:

"If any force generates a motion, a double force will generate double the motion, a triple force triple the motion, whether that force be impressed altogether and at once, or gradually and successively. And this motion (being always directed the fame way with the generating force) if the body moved before, is added to or subtracted from the former motion, according as they directly conspire with or are directly contrary to each other; or obliquely joined, when they are oblique, so as to produce a new motion compounded from the determination of both".

It implies as force is applied its motion (velocity) increases.

**Law II:** The alteration of motion is ever proportional to the motive force impress'd; and is made in the direction of the right line in which that force is impress'd.

### *Explanation to Law II*

Here, the literary meaning of the word alteration is *'change'*

Here, the literary meaning of word motion is *'A natural event that involves a change in the position or location of something.'*

The alteration of motion means 'change in movement' or change in velocity or

$$\text{terminal velocity} - \text{initial velocity or } v - u$$

The word 'alteration' never means derivative with respect to time, i.e. $\frac{d}{dt}$. Also, Newton described three laws of motion in his Philosophiae Naturalis Principia Mathematica originally written in Latin. It was translated into English as The Mathematical Principles of Natural of Philosophy by Andrew Motte and published in 1729 two years after the death of Sir Isaac Newton. The three laws of motion are discussed on pages 19–20. But Newton did not write the word 'alteration' as a derivative with respect to time i.e. $d/dt$. Did Newton write this notation at some other stage? Newton published three editions of the Principia, in 1687, 1713 and 1726. The first English translation of the Principia was published in 1729 and is based on the 1726 edition of the Principia. The third and final edition of the Principia was published 39years after the first edition. In these 39 years Newton did not write the time derivative $d/dt$ in second law of motion. Had Newton any intention for doing so the 'rate of change of' would have been added to it. As Newton did not want to do so, thus the second law was not altered in these 39 years. Needless to mention, he was the inventor of calculus. So there is no possibility that Newton had added to this afterwards. Thus the law is absolute in this form. In the Principia Newton never used the term $d/dt$ for alteration. Thus, to draw final conclusions the related researchers of Newton are to be studied. When Newton defined the second law of motion in the Principia, the mathematical equation $F = ma$ or $F = Kma$ was not quoted at all. When were these equations quoted? Why is this equation associated with Newton when it was not given by him? It has to be assessed whether Newton or other scientists quoted these equations involving time derivatives. The answer to this question can be found by studying the relevant literature. But alteration in motion cannot be the rate of change of motion.

A body is said to be in motion if it changes its position with regard to other bodies.

For example a moving bus is said to be in motion when it changes its velocity. Thus, the change in motion means the change in velocity $(v_2 - v_1)$

$$\text{Alteration in motion} = v_2 - v_1 = \Delta v \qquad (8.6)$$

Thus, according to the Principia, $F \propto (v - u)$ is Newton's second law of motion, and thus $F = Kma$ may be termed the second law of motion.

### *7.1. Mathematical expression of Newton's second law of motion*

"The alteration of motion is ever proportional to the motive force impress'd; and is made in the direction of the right line in which that force is impress'd."

$$\text{Thus, alteration in motion} = v_2 - v_1 \qquad (8.6)$$

$$\text{alteration in motion} \propto \text{impressed force}$$

or $$\text{Impressed force} \propto (v_2 - v_1)$$

or $$F = C(v - u) \qquad (8.7)$$

where $C$ is the coefficient of proportionality which is determined experimentally. It is constant if found the same in all experiments. $u$ and $v$ are initial and final velocities. The coefficient $C$ has the dimensions $MT^{-1}$, according to Eq. (8.7).

### 7.2. Second law of motion as taught in textbooks is not given by Newton

Newton proposed the second law of motion in the Principia (hence called Newton's second law of motion), i.e. $F = C(v - u)$. However, the prevalent second law of motion as given in the literature has not been given by Newton (hence called the second law of motion), i.e. $F = m(v - u)/t$.

Second law of motion:

'The rate of change of momentum is ever proportional to the motive force impress'd; and is made in the direction of the right line in which that force is impress'd.'

In terms of mathematical equations

$$F = Kma = Km\,\frac{(v-u)}{(t_2-t_1)} \qquad (8.8)$$

where $K$ is the constant of proportionality, $m$ is mass and $a$ is acceleration. $v$ terminal velocity, $u$ initial velocity, $t_2$ final time and $t_1$ is initial time. In the existing literature Eq. (8.8) is known as Newton's second law of motion but it is not so it may be called the second law of motion only.

Newton's second law of motion is mathematically given by Eq. (8.7), $F = C(v - u)$. Unfortunately, the exact form of Newton's second law of motion is neglected by scientists. It is not clear when, how and why the second law of motion $F = Kma$ replaced Newton's second law of motion

$$F = C(v - u) \qquad (8.7)$$

Newton did not mean motion as a momentum; it is evident from Definition II on page 4 of the Principia. Thus, Newton's second law of motion does not mean $F = ma$ [2]. Both Newton's second law of motion i.e. $F = C(v - u)$, and the second law of motion $F = m(v - u)/t$ reduce to the first law of motion ($u = v$) under the condition when no external force acts ($F = 0$).

## *7.3. Definition II in the Principia Book 1*

Newton had given various definitions as he used in Book I of the Principia [1] starting from page 2–8.

The quantity of motion is defined in Definition II: 'The quantity of motion is a measure of the same, arising from velocity and quantity of matter conjunctly (involving the joint activity of two or more)'.

It is obvious that the quantity of motion involves velocity and mass (quantity of matter) collectively. Thus Newton defined 'quantity of motion' as we understand momentum ($mv$) nowadays.

The body is said to be in motion if it changes its position with respect to other bodies or possesses velocity. Motion means a change in position or a change in distance.

Thus 'quantity of motion' and motion are different quantities. The motion simply represented by velocity.

Even now the total quantity of motion contained is called the momentum. Simply motion is not momentum, it is velocity. Thus Newton meant the momentum as 'the quantity of motion'. Hence by motion Newton simply meant the velocity and the alteration in velocity meant $v_2 - v_1 = \Delta v$

## *7.4. Newton discovered the calculus and did not use derivative of time, i.e. d/dt*

In standard textbooks the theme of the second law of motion is: "The rate of change of momentum is ever proportional to the motive force impress'd; and is made in the direction of the right line in which that force is impress'd."

This law was not stated by Newton. The word 'alteration' never means the derivative with respect to time, i.e. $d/dt$. Newton did not write the word 'alteration' as the derivative with respect to time, i.e. $d/dt$. Did Newton write this notation at some other stage? It is out of question as Newton published the last edition of the Principia in the Latin in 1726. Had Newton altered the law it would have appeared in the last edition of the Principia. The last edition of the Principia appeared in 1726 39 years after the first one. Thus, in this period

Newton did not change the law, as he was fully satisfied with this. When Newton associated the force with first law of motion it became the second law of motion. Both the laws involve the word motion which means velocity. Also, if the mathematical equation of the law $F = C(v - u)$, is interpreted then $u = v$, which implies the first law. When Newton defined the second law of motion in the Principia, then even the mathematical equation $F = ma$ or $F = Kma$ was not quoted at all. When were these equations quoted? It has to be assessed whether one or more scientists quoted these equations involving time derivatives. The answer to this question can be found by studying the relevant literature.

## *7.5. Who defined Newton's second of motion in the form F = kma?*

The strangest aspect of Newton's law is that its definition in the textbooks has changed from the definition as given in the Principia. Newton published the three editions of the Principia in 1687, 1913 and 1726 in Latin. In every edition there were some additions and alterations, but the second law of motion remained the same in the Principia. As already mentioned, the third edition of the Principia was published 39 years after the first edition. Had Newton mean some different form of the second law or different explanation, he would have published the same in the second and third editions. The first English translation of the Principia was published in 1729 which also quotes the second law of motion as stated in the Latin. Newton died in 1727, so the definition could not have been changed after his death. The first and third laws of motion are the same in the textbooks as given by Newton in the Latin and the first translation in 1727 on page 19. Why is there only a change in the second law of motion and not in the first and third laws of motion in the Principia? How the status of Newton's Second Law of Motion is changed, i.e. $F = ma$ is called Newton's second law of motion whereas actually Newton stated the second law of motion in the Principia as $F \propto (v_2 - v_1)$ ? How te status of Newton's second law of motion $F \propto (v_2 - v_1)$ is changed to $F = m(v-u)/t$? What has been the motive of changing the status of Newton's second law of motion? The name of the scientist or mathematician or philosopher who changed the status of Newton's second of motion in the Principia is not known. All these questions need scientific consensus. The laws of science are neither snatched nor gifted.

So it is ethically, scientifically and logically unjustified to change the definition of the law without the consent of the author and attach any law to the author. It appears the scientists who changed the law were not aware of the Principia. But it is the strangest fact that the second law of motion, as we teach today, is different from what Newton had given in the Principia in original Latin and English translation. This aspect is also pointed out in the existing literature by Bruce Pourciau as [2,3]:

"When we call the equation $F = ma$ 'Newton's second law'," how much historical truth lies behind us? Many textbooks on introductory physics and classical mechanics claim that the Principia's second law becomes $F = ma$, once Newton's vocabulary has been translated into more familiar terms. But there is nothing in the Principia's second law about acceleration and nothing about a rate of change. If the Principia's second law does not assert $F = ma$, what does it assert?" This is an independent conclusion and also confirms the same that Newton did not discover $F = ma$ or $F = Kma$. When Newton stated the second law of motion in the Principia [1] on page 19, he did not use or mention the derivative $d/dt$, and also $F = Kma$ or $F = ma$ was not quoted. It is pertinent to mention that Newton was also the inventor of the calculus. Thus, Newton knew about the functions of derivatives as he was the discoverer of the calculus. So if he had to express the law in terms of d/dt then he would have done it. Moreover the third edition of the Principia was published 39 years after the first edition, so Newton had ample time to change the law if he wanted. So Newton's second law of motion meant that the change in motion (velocity) is proportional to force:

$$F = C(v - u) \tag{8.7}$$

where $C$ is the coefficient of proportionality which is determined experimentally. There is no clear evidence in the existing literature (as old literature is not easily available) as to who proposed the equation $F = ma$. The comparison of Newton's second law of motion and the second law of motion is shown in Table II.

### *7.6. Are p = mv and F = ma analogous ?*

Apparently, the equation $F = ma$ may have been given in the following way analogous to the momentum. The velocity and acceleration are related quantities

(i) The product of mass and velocity ($v$) is the momentum

$$P = mv \tag{8.4}$$

(ii) What would be the product of mass and acceleration ($a$) i.e. $ma$? Apparently, on the analogous basis scientists, philosophers or mathematicians coined '$ma$' as the force, i.e.

$$F = ma \tag{8.5}$$

As $mv$ is the momentum, $ma$ is the force. Thus, the force was related to Newton's second law of motion. But it is only guesswork

**Table II.** Comparison of Newton's second law of motion (Newton defined it in the Principia) and second law of motion (the law which Newton did not define)

| Sr. No | Newton's second law of motion $F = C(v - u)$ | Second law of motion $F = m(v - u)/t$ |
|---|---|---|
| 1 | Definition<br>The alteration of motion is ever proportional to the motive force impress'd. | Definition<br>The rate of change of momentum is ever proportional to the motive force impress'd. |
| 2 | Origin<br>The Principia, 1687 | Origin<br>Unknown, year? |
| 3 | Inventor<br>Sir Isaac Newton | Inventor<br>Nameless |
| 4 | Mathematical equation<br>$F = C\,(v - u)$<br>$C$ is the coefficient of proportionality | Mathematical equation<br>$F = Km\,\dfrac{(v-u)}{(t_2 - t_1)}$ |
| 5 | If $F = 0$, then<br>$v = u$, i.e. first law of motion | If $F = 0$, then<br>$v = u$, i.e. first law of motion |
| 6 | Peculiar fact<br>Newton discovered but not associated with Newton | Peculiar fact<br>Newton did not discover it but it is associated with Newton |
| 7 | Inertial mass<br>Force not associated with inertial mass. | Inertial mass<br>$m = \dfrac{F}{a} = \dfrac{0}{0}$ |
| 8 | Future<br>Coefficient of proportionality $C$ must be determined | Who discovered $F = ma$?<br>When $F = ma$ is invented?<br>How it crept into the Principia?<br>How it is called Newton's second law of motion? |
| 9 | Morale<br>Science is infinite | Equation of mass must be such that denominator and numerator both are non-zero i.e. positive. |

as there are no direct evidence to this conclusion. Further, the product of mass and distance can be speculated.

## 7.7. $F = ma$ does not follow from Newton's second law of motion

From the definition of the second law of motion

Alteration in motion $\propto$ Motive force impressed

$$F = C(v - u) \tag{8.7}$$

In no way Sir Isaac Newton wrote (while quoting second law of motion in the Principia)

$$\text{alteration in motion} = \text{rate of change in momentum } \left(\frac{dp}{dt}\right) \tag{8.9}$$

When Newton stated the second law of motion in the Principia on page 19, the derivative $d/dt$ was not used or mentioned, and even $F = Kma$ or $F = ma$ was not quoted. It is pertinent to mention that Newton was also inventor of calculus. Thus legend knew about functions of derivatives. Anyhow, let us move with the text quoted in the numerous publications. Thus,

$$\frac{dp}{dt} \propto \text{Motive Force Impressed}$$

To remove the sign of proportionality, the coefficient of proportionality is introduced and its value is determined experimentally. If its value remains the same in all cases then the coefficient of proportionality is called 'the constant of proportionality.' In the existing literature there is no evidence that any attempt was made to measure the value of $K$ experimentally.

$$\frac{dp}{dt} = k \text{ Motive force impressed}$$

$$\frac{dp}{dt} = k\,F$$

or

$$F = \frac{1}{k}\frac{dp}{dt}$$

or

$$kF = ma \tag{8.10}$$

or

$$F = Kma$$

It has already been mentioned that when Newton defined the second law of motion, $F = Kma$ or $F = ma$ were not written hence the question of determination of $K$ did not exist. The value of $K$ may have been assessed at a later stage. The English scientist Henry Cavendish experimentally measured the value of the universal gravitational constant $G$ in 1798, however the law of gravitation was stated in 1687. The equivalent method must be used for the measurement of $K$.

### *7.8. The value of the coefficient of proportionality K in F = Kma is determined assuming values of all variables are equal to unity*

In the existing literature the value of $K$ is regarded as unity in all the cases (various bodies, different accelerations, in various media) in the following way. It is assumed that if a body has unit mass ($m = 1$) and unit acceleration ($a = 1$) such that the force possessed by it is also unit ($F = 1$) then the value of $K$ is unity. Now, Eq. (8.8) becomes

$$m = 1,\ a = 1,\ F = 1$$

$$1 = K \times 1 \times 1$$

or

$$K = 1 \tag{8.11}$$

In the motion of bodies resistive forces (atmospheric, frictional and gravitational, etc), play significant roles but this fact is not taken into account. Who was the scientist who measured the value of $K$ as unity? Is it an established method for the measurement of the coefficient of proportionality? If so, why it is not applied in other cases? Why is the equation $F = Kma$ regarded as an exception in this regard? Newton did not measure the value of $K$ as he had not given equation $F = Kma$. The value of $K$ can not be measured by the method of arbitrary assumption that assumes all parameters or variables as unity and then calculate $K$ as unity.

The answer to this question can only be found by studying the related literature. The second law of motion is applicable for all bodies of different masses in ideal and practical media. If the resistive forces are infinitesimally negligible, then with a smaller force the body can move a larger distance. Secondly, if the resistive forces in the medium are maximum, then the body will travel the shortest distance. $F = Kma$ is applicable to various masses, some of them are

$$m_e \text{ (mass of electron)} = 9.1 \times 10^{-27} \text{ kg}$$

mass of man = 78 kg
mass of elephant = 300 kg
mass of the Sun = $1.9891 \times 10^{30}$ kg
mass of Jupiter = $1.89813 \times 10^{27}$ kg

Similarly, the second law of motion, i.e. $F = Kma$, is applicable to various accelerations, however, the value of $K$ is only determined for the unit acceleration and the body of unit mass.

The methods of assumption are not scientifically valid methods for the determination of coefficients of proportionality in science or of any other parameter. Now, if this method is applied to all equations in physics, then all values of the coefficients of proportionality in physics will be regarded as unity. The coefficient of proportionality $K$ in $F = Kma$ is determined by assuming the values of all variables as unity. Thus, the value of $K$ turns out to be unity. If this method is used, then the value of the universal gravitational constant will also turn out to be unity. In Newton's second law of motion and other laws the resistive force of the system also plays a significant role. These forces have to be accounted for complete understanding of the problem. We will discuss its implications in measurement of the universal gravitational constant $G$ in this case and in reference to other coefficients.

## 8.0. Coefficient of proportionality *G* in law of gravitation

The coefficient of proportionality is determined experimentally and if its value is measured the same in all cases then it is called the constant of proportionality, otherwise the coefficient of proportionality.

**Example of constant:** Newton stated the universal law of gravitation as: 'every body of the universe attracts every other body towards its centre, with a force which is proportional to the product of masses of bodies and inversely proportional to the square of the distance between them.'

Mathematically,

$$F = G\frac{mM}{r^2} \tag{8.12}$$

where $F$ is the force of attraction between two bodies. $M$ and $m$ are the masses of two bodies and $r$ is the distance between their centres. $G$ is the universal gravitation constant. The value of $G$ is called the universal gravitational constant and its value is experimentally measured. For the first time it was measured in 1798 by the English experimentalist Henry Cavendish at Cambridge University. The value of $G$ is the universal gravitational constant, $6.67300 \times 10^{-11}$ m$^3$ kg$^{-1}$ s$^{-2}$.

**Coefficient of proportionality:** Both $G$ and $K$ have similar origins (originate from the equation of force), now the value of $G$ is experimentally measured whereas that of $K$ is regarded as unity without logic and experiments. The coefficient of proportionality always depends upon *experimental conditions.* The gravitational constant $G$ is used in measuring the mass of the Earth and the distance between the Earth and the Sun. When $G$ and $K$ have similar origins then why are the methods of determination of $G$ and $K$ different, in fact the method of measurement of the coefficient of proportionality must be the same in all cases. Thus, it is scientific logic that the value of $K$ must be determined experimentally as in the case of other laws in physics.

The value of $K$ is determined as unity arbitrarily. Thus, values of mass and acceleration are assumed arbitrarily to be unity. Under this assumption ($m = 1$, $a = 1$, $F = 1$) the value of $K$ also turns out to be unity. Equation (8.8) also gives the force, hence the same method (assuming all values of variables in the equation as unity, except the coefficient of proportionality) can be applied here.

In Eq. (8.8) the force will be unity (say 1 N) if

$$m = 1,\ M = 1,\ r = 1 \text{ and } F = 1$$

$$1 = G\ \frac{1 \times 1}{1}$$

or

$$G = 1 \tag{8.13}$$

Then the forces of attraction between various bodies will vary. The acceleration due to gravity will increase many times on the Earth. Let us theoretically consider the following numerical problem.

## *8.1. Arbitrary method of determination of G is not justified*

Assuming the Earth to be a sphere with a radius of 6400 km and a density of 5.5 g/cc, we find the value of $g$ on the surface of the Earth; $G = 1$ N m$^2$kg$^{-2}$, as determined in Eq. (8.13).

**Solution:** Here radius of the Earth = $6400 \times 10^3$ *m* = $6.4 \times 10^6$ m

Density = 5.5 g/cc = $5.5 \times 10^3$ kg/m$^3$

$$g = \frac{GM}{R^2} = \frac{G}{R^2} \times \frac{4}{3}\pi R^3 \times D$$

$$\frac{4}{3} = \pi GRD$$

$$= \frac{4}{3} \times \frac{22}{7} \times 1 \times 6400 \times 10^3 \times 5.5 \times 10^3$$

$$= 1.475 \times 10^{11} \text{ m/s}^2 \tag{8.14}$$

which is not justified. The value of $g$ is measured equal to 9.8 m/s$^2$ by various methods. If the value of $G$ is measured by an arbitrary method as above, then value of $g$ turns out to be $1.475 \times 10^{11}$ m/s$^2$, which is never justified. Hence here the fault is in the method of estimation of $G$ which is measured arbitrarily in Eq. (8.13), thus the measurement of $K$ in $F = Kma$ is also arbitrary. Thus, the value of $G$ has to be measured experimentally as done by Henry Cavendish and in this case the value of $g$ comes out to be 9.8 m/s$^2$ which is supported by experiments.

## *8.2. The value of* the *coefficient of proportionality $\eta$ can be measured by this method*

As in the equation of gravitational attraction, i.e. Eq. (8.12), the value of $G$ is measured experimentally, likewise the value of $K$ in $F = Kma$ has to be measured experimentally. The value of the constant or coefficient of proportionality depends upon the involved parameters. If the coefficients of proportionality are to be taken as unity as in Eq. (8.11), then the coefficient of viscosity, the coefficient of expansion(linear, superficial and cubical), the coefficient of thermal conductivity, the Hubble coefficient, the coefficient in Coulomb's law, the modulus of elasticity, etc., are also unity. There is no special status to $F = Kma$ in any way in science. There has to be same method for determination for all coefficients of proportionality. Consider one example of the viscous force

$$F = \eta\, A \frac{dv}{dx} \tag{8.15}$$

We have the equation for the second law of motion

$$F = Kma \tag{8.8}$$

In Eq. (8.8) the values of $m$, $a$ and $F$ are taken as unity, then

$$K = 1$$

$dv$, $dx$, $A$ and $F$ can be unity (as in the case of the second law of motion), i.e. $dv = 1$, $dx = 1$, $A = 1$, $F = 1$ (as in $F = Kma$, $F = 1$, $m = 1$, $a = 1$)

$$1 = \eta \times 1 \times \frac{1}{1}$$

or

$$\eta = 1 \tag{8.16}$$

However, the value of the coefficients of viscosities for different liquids are different and experimentally measured. They are not measured by assumptions but by actual measurements using gradually refined and improved methods.

| Liquid | Coefficient of viscosity |
|---|---|
| Ethyl alcohol | $1.32 \times 10^{-4}$ poise |
| Mercury | $15.5 \times 10^{-3}$ poise |
| Water | $1.01 \times 10^{-2}$ poise |
| Glycerine | 10.69 poise |

There are many examples in physics that the coefficients of viscosity are experimentally determined. But it is not so in the case of the value of $K$ in the second law of motion where the resistive forces of the system play a significant role. If experiments are conducted for the measurement of other coefficients of proportionality, then why not for $F = Kma$ ? Moreover, science is an expanding body, not closed. Newton never mentioned about $F = Kma$ or $F = ma$ in the Principia (in any of three volumes, i.e. in Book I, Book II and Book III), hence the measurement of $K$ was out of question at that stage. The three laws of motion are mentioned in Book I on page 19 of the Principia, and Newton never mentioned $F = Kma$ at any stage. So the equation $F = Kma$ was not derived by Newton.

Also, the explanation for the three laws is given in a single paragraph. By and large the Principia deals primarily with geometrical analysis.

### *8.3. Why special status for F = Kma?*

The equation $F = Kma$ has a very special status on two counts:

(i) It is called Newton's second law of motion (but in the discussion it is mentioned only as the second law of motion). However, it is not mentioned at all in any of the editions of the Principia, i.e. 1687, 1713 and 1726. Also there is no mention in the literature who and when discovered $F = Kma$ which replaced $F = C(v - u)$, and further how it is credited to Newton. It can be speculated whether this equation existed before Newton or not. The concept of acceleration existed before Newton. At that time there were no science journals, magazines etc. as we have now. Despite this, it is called Newton's second law of motion; however, Newton neither discovered it nor published it in the Principia.

(ii) The method of measurement of $K$ in $F = Kma$ is arbitrary, i.e. values of all variables are regarded as unity, i.e. $m = 1$, $a = 1$ and $F = 1$. If the same method is applied to the coefficient of proportionality then the universal constant $G$ becomes unity. In the motion of bodies the resistive forces play a significant role and this has to be taken into account, so that such forces cannot be neglected. Therefore, the same method must be applied in the determination of $K$ (and other coefficients); thus, the coefficient of proportionality must be measured experimentally.

## 9.0. Newton's Principia and the third law of motion

Newton's most famous book popularly known as the Principia was written originally in Latin as Philosophiæ Naturalis Principia Mathematica in 1687. It was translated to English by Andrew Motte in 1729 as 'Mathematical Principles of Natural Philosophy'. Two full English translations of Newton's 'Principia' have appeared, both based on Newton's 3rd edition of 1726. Newton's three laws of motion in original Latin and English are given below.

Newton's third law of motion as stated in original Latin

***Lex III:*** *Actioni contrariam semper et æqualem esse reactionem: sive corporum duorum actiones in se mutuo semper esse æquales et in partes contrarias dirigi.*

When translated to English:

**Law III:** To every action there is always an equal and opposite reaction: or the forces of two bodies on each other are always equal and are directed in opposite directions.

Newton explained the law in the following way in the Principia. Whatever draws or presses another is as much drawn or pressed by that other. If you press a stone with your finger, the finger is also pressed by the stone. If a horse draws a stone tied to a rope, the horse (if I may so say) will be equally drawn back towards the stone.

### *9.1. Action and reaction are not always equal and opposite*

Consider a body $A$ of weight $W$ is lying over the surface of the other body $B$. The body $A$ exerts force $W$ on the other body $B$. Thus,

$$\text{Action of body } A = W \tag{8.17}$$

According to Newton's third law of motion the body $B$ exerts the equal and opposite reaction $R$ to the body $A$

$$\text{Reaction of body } B = R \tag{8.18}$$

Now both the acting and reacting bodies remain at respective positions, thus

$$\text{Action of body } A = -\text{ Reaction of body } B$$

or

$$R = -W \tag{8.19}$$

If in general body $A$ exerts force $F_{AB}$ on body $B$, then body $B$ exerts force $F_{BA}$ on body $A$ and according to Newton's third law of motion

$$F_{AB} = -F_{BA} \tag{8.20}$$

Thus, action and reaction are equal and opposite.

Also, there are some qualitative examples for understanding the third law of motion. For example, a swimmer pushes water backward and moves in the forward direction. It easier to walk on the Earth as we get sufficient reaction. On the other hand, it is difficult to walk on the sand as the adequate reaction is not obtained. A ball is hit on the wall then as a reaction it rebounds. In elastic collisions if the target is of a sufficiently large mass, then as a reaction the projectile rebounds. It can be a general law if it is quantitatively confirmed in all cases where it is applicable.

### *9.2. Following example is neglected*

Consider a boy is standing in front of a concrete wall, holding a rubber ball and a cloth ball in the hands. Let the wall is at the distance of 5 m from the boy.

**First case:** Let the boy throws the rubber ball at the wall with force *F*. The ball strikes the wall, and comes back to the boy, i.e. travelling 5 m. Now action and reaction can be understood as

Action: the boy pushes the ball towards the wall
Reaction: the wall pushes the ball towards the boy.

As the boy pushed the ball from a distance of 5 m and the wall rebounds the ball to distance of 5 m the action and reaction are equal and opposite.

Boy pushes the ball through 5 m
= Wall rebounds the ball through 5 m

$$\text{Action} = -\text{Reaction} \tag{8.21}$$

The *–ve* sign indicates that after the reaction the ball moves in the opposite direction. In this case Newton's third law of motion completely obeyed.

**Second case:** Let the boy throws the cloth ball at the wall which is at a distance of 5 *m*.

Let the boy throws the cloth wall with same force *F*. The ball strikes the wall and rebounds to only 3 m. Now action and reaction can be understood as

Action: the boy pushes the ball towards the wall
Reaction: the wall pushes the ball towards the boy

The boy pushed the cloth ball from a distance of 5 m and the wall rebounds to a distance of 3 m.

Thus boy pushes the ball towards the wall through 5*m*
The wall rebounds the ball through 3*m*

$$\text{Action} \neq -\text{Reaction}$$

So action and reaction are opposite but not equal in the case of the cloth ball. In addition, the quantitative confirmation of the mathematical equations based upon elastic collisions (when the mass of the target is sufficiently larger than the projectile) will also lead to concrete results in this regard. The experimental verification of such equations includes the measurement of rebound velocities of projectiles. There can be many such examples. It is just possible that the characteristics of the ball and the wall are such that the ball may rebound more than 5 m.

**Conclusions:** The reaction also depends on various characteristics of the body or the ball (cloth, rubber) and the wall (concrete or wooden). Thus, the reaction depends on various characteristics of the system. In the case of elastic collisions, the natures of projectiles, targets and surface plays a significant role. There can be many such examples. The quantitative confirmation of the mathematical equations based on elastic collisions (when the mass of the target is sufficiently larger than that of the projectile) will lead to concrete results in this regard.

According to the definition of the third law of motion, the law is absolute or unrestricted, i.e. the forces of two bodies on each other are equal and are directed in opposite directions. However, here the forces acting on the ball and the wall are opposite but not equal. The inherent characteristics of the process play a significant role. Thus, the third law of motion is generalized.

**Generalization:** The above discussion allows us to reassess Newton's third law of motion as: 'to every action there is always an opposite reaction which may or may not be equal depending on the characteristics of the system'.

$$\text{Reaction} \propto \text{Action or Reaction} = -k \text{ Action} \tag{8.22}$$

Here $k$ is the coefficient of proportionality which depends upon the characteristic conditions of the process. If $k = 1$, then

$$\text{Reaction} = -\text{Action} \tag{8.23}$$

Thus, the third law of motion is generalized in the case of macroscopic bodies. Some suggested experiments will further be useful in understanding the law.

### *9.3. Second law of motion is the real law of motion*

The second law of motion is called the real law of motion, Newton's first law of motion can be obtained from the second law of motion. Also Newton's third law of motion can be derived from the second law of motion. Both the first and third law of motion are derived from the second law of motion under the condition if no external force ($F = 0$, $a = 0$, $u = v$) acts on the system. Basically, the Newton's first law of motion is a statement and the second law of motion is its mathematical form ($F = \frac{1}{k}\frac{dp}{dt}$ or $F = Kma$). The first law of motion is the other form of Galileo's law of inertia (in the absence of external force $F = 0$, the body moves with uniform velocity, i.e. $u = v$). Newton has adopted the law of inertia as the first law of motion, whereas the origin of the third law of motion is independent of the first and second law of motion.

**(a) Newton's first law of motion follows from the second law of motion:**

According to the second law of motion, the external force $F$ applied to the body and the acceleration $a$ produced are related to each other by the relation

$$F = Kma$$

If no external force acts on the body, i.e. $F = 0$, then

$$Kma = 0$$

As mass $m$ and $K$ cannot be zero, hence $a = 0$
Now the acceleration is defined as the rate of change of the velocity of the body

$$a = \frac{(v-u)}{t}$$

As $a = 0$, then

$$0 = \frac{(v-u)}{t} \text{ or } v = u \tag{8.24}$$

Terminal velocity = Initial velocity (if external force $F = 0$) which is nothing but Newton's first law of motion.

'The body continues with uniform velocity if no external force acts on it.'

If the body is once set in motion it maintains uniform velocity ($u = v$) which is true if resistive forces (atmospheric force, frictional force, gravitation force, etc.) are completely absent. Under similar perception the law of inertia was given by Galileo before Sir Isaac Newton.

**(b) Newton's third law of motion follows from the second law of motion.**

Consider a system of two bodies $A$ and $B$ mutually interacting with each other, such that no external forces act on the system. Suppose that body $A$ exerts force $F_{AB}$ (action) on body $B$; and body $B$ exerts force $F_{BA}$ (reaction) on body $A$.

If $\frac{dp_1}{dt}$ and $\frac{dp_2}{dt}$ are the rates of change of momentum of bodies $A$ and $B$ respectively due to forces $F_{AB}$ and $F_{BA}$ acting on them then

$$F_{BA} = \frac{dp_1}{dt}$$

$$F_{AB} = \frac{dp_2}{dt}$$

$$F_{BA} + F_{AB} = \frac{dp_1}{dt} + \frac{dp_2}{dt} = \frac{d(p_1 + p_2)}{dt} \quad (8.25)$$

Since no external forces act on the system, the total momentum of the system must be constant, i.e. the rate of change of the momentum must be zero. Thus,

$$\frac{d(p_1 + p_2)}{dt} = 0$$

Hence, Eq. (8.25) becomes

$$F_{BA} + F_{AB} = 0$$

or

$$F_{BA} = -F_{AB} \quad (8.26)$$

i.e. action and reaction are equal and opposite (if no external forces act on the system).

Hence Newton's third law of motion follows from the second law of motion when no external forces act on the system.

From the above discussion we conclude that:

> Newton's first law of motion follows from the second law of motion.
> Newton's third law of motion follows from the second law of motion.

Hence the second law of motion is the central law of motion.

It must be noted that Newton's 2nd and 3rd laws of motion follow from $F = ma$ (the mathematical form of the second law of motion) if no external force ($F = 0$) acts on the system.

## 10.0. Prediction of undefined inertial mass from equation $F = ma$ or $m = \frac{F}{a}$

**Inertial mass:** When a body is in translation under the effect of external force other than gravity, the mass of the body measured is called its inertial mass.

Let force $F$ produces the acceleration $a$ in the body of mass $m$, then mathematically the inertial mass is given by

$$m = \frac{F}{a} \tag{8.27}$$

### *10.1. Inertial mass under the conditions when Newton's second law of motion reduces to first law of motion ($a = 0$, $u = v$, $F = 0$).*

We have just proved that the second law of motion implies that when no external force ($F = 0$) acts on the system, then the body moves with uniform velocity ($u = v$, $a = 0$). In other words, the body continues in the state of uniform motion when no external force acts on it (just the law of inertia). It is the second part of the first law of motion. Thus, under this condition,

$$F = 0, \quad a = 0$$

Now, the equation for inertial mass, i.e. Eq. (8.5), becomes

$$m = \frac{F}{a} = \frac{0}{0}$$

It is reiterated that under this condition ($F = 0$, $a = 0$), the second law of motion reduces to the first law of motion. But under this condition ($F = 0$, $a = 0$) inertial mass becomes UNDEFINED, hence all other interpretations are MEANINGLESS. This aspect is not discussed in the existing literature.

Dimensions of LHS = $M$
Dimensions of RHS = undefined

Units of LHS = kg (SI system)
Units of RHS = undefined (in every system)

Although division by zero is not allowed yet, it naturally follows from the mathematical equation based upon the second law of motion.

In addition, in this case the numerator also becomes zero under feasible conditions, i.e. when $F = 0$, $a = 0$ ($u = v$, unaccelerated motion). In this case, the second law of motion reduces to the first law of motion. The mass of the body is always positive and all physical laws are always meant for bodies of non-zero mass. Thus, the equations must give non-zero and exact values of mass. If an equation gives an inconsistent value of mass (like indeterminate or zero for non-zero mass), then that equation needs to be changed. Purposely we can speculate that the denominator must be the sum of velocities rather than the difference of the two [4]. If some equation gives an indeterminate value of mass and the consequences are not discussed then it is unscientific. Also it is NOT limitation of the legend as Newton did not propose the equation $F = Kma$. It is a mystery. Now the old scientific literature must be scrutinized to find out who derived $F = Kma$ and how it replaced in Newton's second law of motion $F = C(v - u)$. Why is the scientific audience silent on this sensitive issue? The first and third laws as we teach today all over the world are the same as given by Newton in the Principia. The net force also becomes zero ($F = 0$) if a number of competing forces act on the system.

The indeterminate form is also obtained in critical analysis of the Archimedes principle. This aspect is completely discussed in Chapter 5 as

$$V = \frac{\left((V+v)D_w - D_w V - vD_w\right)}{\left(D_w - D_w\right)} = \frac{\left(vD_w - vD_w\right)}{\left(D_w - D_w\right)} = \frac{0}{0} \tag{5.27}$$

or the volume of platinum, silver, cork, metal and air $= \frac{0}{0}$

It can be explained by generalizing the principle and we get

$$V = \frac{\left(f(V+v)D_w - VD_w - fvD_w\right)}{\left(fD_w - D_m\right)} = \frac{(f-1)D_w V}{(f-1)D_w} = V \tag{5.37}$$

Thus the inconsistent results can be easily explained by modifying the equations. The science is like lighting one lamp from another.

### *10.2. Core area of applicability of Newton's first and second laws*

Newton's first law of motion implies that if a body once set in uniform motion will continue to move with uniform motion if no external forces ($F$ = 0) act on the system. If resistive forces are present then the body comes to rest. So the body maintains uniform motion in the absence of resistive forces. (If the motion of the body is to be maintained in the resistive systems then external forces are required, which is stated by Aristotle). Also a force is required to set the body in motion, it cannot move on its own.

The second law of motion reduces to the first law of motion if no external forces ($F$ = 0) act on the system of the body and the surface. In this case the body moves with uniform velocity eternally. Mathematically, this situation is described by the equation,

$$F = ma$$

where $F$ is the impressed force, $m$ is mass and $a$ is acceleration

On the Earth frictional force, atmospheric force, gravitational force are always present due to these the body slows down and comes to rest. The body will perfectly move with uniform velocity if the external forces, i.e. frictional force, atmospheric force, gravitational force, are completely absent. But these forces cannot be nullified on the Earth. Thus, bodies gradually stop on the surface of the Earth. However, the uniform motion can be prolonged if the surface on which the body moves is smooth and free from resistive forces.

## *10.3. The facts Newton did not explain*

It is true that a body moving with uniform velocity comes to rest when resistive forces (atmospheric, frictional, gravitational forces etc.) act on the Earth. No force is required to maintain the uniform motion of the body if resistive forces are not present. But how is the body is set in motion with initial velocity initially? The motion is caused by the external force. Thus a force is required for initial motion? According to Newton, the body preserves its uniform motion, i.e. $u = v$, it is true under the conditions when resistive forces are not present (the body is initially in motion). How is the motion of the body maintained if resistive forces are present? Newton's first law of motion is another or extended form of inertia, the concept Galileo developed prior to the discovery of gravitation.

The Greek philosopher Aristotle (364–322 BC) taught that a force is continuously required to maintain the motion of the body. It is even now true if resistive forces (atmospheric, frictional, gravitational forces etc.) are present. Even now it can be justified with the following observations in daily life on ground, water, air, space and atomic scale. Again, the following observations are based upon

(i) To pull a table in a room some external force has to be applied. As soon as external forces ceases to act the table stops.

(ii) If a boat is pushed in water (say in a lake) then it travels some distance and stops. If it is pushed with a greater force it travels a longer distance.

(iii) A toy aeroplane (made up of paper) travels a longer distance if pushed with a greater force.

(iv) The Earth and other planets move around the Sun and the centripetal force is provided by the gravitational force.

(v) Similarly, the centripetal force is provided by the electrostatic force when electrons revolve around the nucleus.

Thus, to maintain the motion of the body obviously an external force is required if the resistive forces are present. This external force is required to overcome the resistive forces, e.g. gravitational force, frictional force and atmospheric force, etc. and provide kinetic energy to the body. The atmospheric and frictional forces can be minimized considerably but it is impossible to eliminate completely the gravitational force and its effect on the Earth. Newton did not try to explain the above situation and confined the studies to the ideal system where the resistive forces are completely absent. This basis was established by Galileo. In such a system (ideal or hypothetical system) if the body is once set in motion then it moves perpetually with uniform velocity. But such a system does not exist on the Earth.

## 11.0. Two types of systems

**(i) Systems free from resistive forces, i.e. ideal system or hypothetical system.** In this system if the body is once set in motion with uniform velocity ($v = u$) then it keeps on moving perpetually with uniform velocity. No force is required to maintain the motion. On the Earth there is no such system where the resistive forces are zero. Gravitational force is always present. So it is a hypothetical or abstraction situation only which does not exist on the Earth. We need to formulate laws not for the hypothetical but for practical system. Newton's first and second laws of motion are meant for this system only. If the resistive forces are zero only then the body once set in motion will maintain its uniform velocity perpetually. While formulating the law of inertia, Galileo confirmed that a ball continues to move with uniform velocity only in the absence of resistive forces. Newton's first law of motion is nothing else but Galileo's law of inertia. Newton's first law of motion and second law of motion are meant for such ideal or hypothetical system only. Such a system cannot be practically visualized on the Earth.

Newton's first law of motion is fully fused with the mathematical form of the second law of motion in such a way that both laws are miscible. Further, due to the simplicity of the mathematical form of the second law of motion ($F = ma$), the first law is not questioned. If there were no 'questions' there would not be any science. Aristotle's laws became dormant when logically questioned by other scientists.

(ii) **Practical systems:** These are the systems in which the resistive forces (atmospheric force, frictional force gravitational force, etc.) are present. Newton stated that the body moves with uniform velocity ($v = u$) in the absence of resistive forces. But these resistive forces are present in various systems hence the body stops after travelling some distance. Thus, Newton admitted that a force is required to maintain the motion of the body if the resistive forces (atmospheric force, frictional force, gravitational force, etc) are present, as the body comes to rest in the presence of the resistive forces. If the resistive force are absent then the body once set in motion will move with uniform velocity. Thus the body cannot move with uniform velocity in every situation (when the resistive forces are present).

Thus, the extrapolated interpretation of Newton's law agrees with Aristotle that a mover is required to move the body in practical systems, as the uniformly moving body comes to halt in the presence of resistive forces. Newton's laws are based upon Galileo's inertia. This inertia is the opposite of Buridan's and Philoponus impetus theory, i.e. a force or its influence is required for the motion of bodies. Aristotle's assertion and Buridan/Philoponus doctrines address the practical systems. Newton did not formulate any mathematical equation for the motion of bodies in the practical systems.

Aristotle proposed a qualitative equation for the motion of bodies as

$$\text{force} = (\text{resistance}) \times (\text{speed})$$

It clearly implies that Aristotle's law is only meant for media when resistive forces are present. Newton's law is meant for cases when resistive forces are completely absent. Hence both the laws are applicable in different conditions and are useful in respective domains (Newton's laws in ideal systems, free from resistive forces, and Aristotle's law in practical systems).

It would be better if scientists formulate a comprehensive equation for the practical system (including atmospheric force, frictional force, gravitational forces, etc.) and this law must reduce to ideal cases (devoid of resistive forces), such as Galileo's law of inertia. The single law must explain the problems in both systems. The prediction of the mathematical equation should be such that a force is required to move the body in practical systems, and the body if once set in motion then must maintain uniform velocity without the force in ideal systems.

## Frequently asked questions

1. **What is Newton's second law of motion? What of the second law of motion?**

   It must be noted that conceptually Newton's second law of motion and the second law of motion are two different laws in every respect i.e. concept, origin and status.

   **Newton's second law of motion**

   The definition of second law of motion from the Principia is: 'the alteration of motion is ever proportional to the motive force impress'd; and is made in the direction of the right line in which that force is impress'd.'

   $$F = C\,(v - u)$$

   where $C$ is the coefficient of proportionality which is determined experimentally. It is regarded as constant if found the same in all experiments. $u$ and $v$ are initial and final velocities. The coefficient $C$ has dimensions $MT^{-1}$.

**Second law of motion**

The rate of change of the momentum is ever proportional to the motive force impress'd; and is made in the direction of the right line in which that force is impress'd.

$$F = Kma$$

where $K$ is the constant of proportionality, $m$ is mass and $a$ is acceleration.

So it must be distinguished which is Newton's second law of motion and which the second law of motion. Newton's second law of motion is the one which Newton published in the Principia. The mathematical equation for this given by Eq. (8.7), i.e. $F = C(v - u)$. The law not given by Newton must not be associated with Newton, hence it is called the second law of motion (not Newton's second law of motion). Thus, the second law of motion is given by Eq. (8.8), i.e. $F = Kma$. Also the first and third laws, which we teach today, are the same as given by Newton in the Principia. Then why only the definition of Newton's second law of motion is changed? There are numerous questions but no answers. It cannot be denied that equation $F = Kma$ may have existed before Newton, as the acceleration existed before Newton. But it is speculation only.

2. **Can Newton's second law of motion, i.e. the alteration of motion is ever proportional to the motive force impressed; and is made in the direction of the right line in which that force is impressed, can lead to $F = ma$ (force = mass × acceleration)**

According to Newton's second law of motion the alteration in motion $(v_2 - v_1) \propto$ force

or force $\propto$ alteration in motion $(v_2 - v_1)$

or
$$F = C\,(v - u) \tag{8.7}$$

where $C$ is the coefficient of proportionality. It is Newton's second law of motion.

It cannot be the second law of motion

$$F = ma \qquad \text{(second law of motion)}$$

or
$$\text{force} \propto a$$

$$\text{force} \propto \frac{(v_2 - v_1)}{(t_2 - t_1)}$$

Thus, Newton's second law of motion, i.e. $F = C(v - u)$, as given in the Principia, does not lead to $F = ma$. In Newton's second law of motion the derivatives of time and acceleration do not exist.

3. **If Newton did not discover equation $F = ma$, then who did and how?**

   **Answer:** There is no mention in the Principia of $F = Kma$ or the definition of the second law of motion as we teach in the standard books now. There is no clear evidence in the literature who has discovered $F = Kma$ or the second law of motion and attributed it to Newton. Anyhow, can we speculate how $F = Kma$ was discovered? The answer lies in the existing literature. It may be possible that $F = ma$ may have existed before Newton, as the acceleration was defined before Newton. The various kinematical equations existed before Newton's time. For example

$$v = u + at$$

$$S = ut + \frac{1}{2}at^2$$

$$v^2 - u^2 = 2aS$$

   Further, the possibility of existence of $F = ma$ like $P = mv$ can not be excluded. All the aspects can be cleared with extensive survey of the literature. Can we speculate a term $mS$ in today's context?

4. **Are Galileo's law of inertia and Newton's first law of motion and second law of motion well established? Is there any scope for further study left in the field?**

   **Answer:** It is true that Galileo's law of inertia and Newton's first law of motion and the second law of motion are well established, but there is a topic which can be further studied. For example, we can develop equations of force for media where resistive forces are present.

   Hence we find that Aristotle (a force is required for the body to move), Philiponus (an arrow moves due to imparted energy), Galileo (the body maintains uniform motion in the absence of resistive forces) or Newton (the same as in case of Galileo) studied problems in individual parts. If one aspect is addressed then the other one is neglected.

   It would be better if scientists formulate a comprehensive equation for practical systems (including atmospheric force, frictional force, gravitational forces, etc.) and this law must reduce to ideal cases (devoid of resistive forces), such as Galileo's law of inertia. The single law must explain the problems in both systems. The prediction of the mathematical equation should be such that a force is required to move the body in the practical system, and the body if once set in motion then must maintain uniform velocity without any force in the ideal system.

## References


1. Newton, L., *Mathematical Principles of Natural Philosophy,* pp. 19–20, London, 1727, translated by Andrew Motte from the *Latin*.
2. Pourciau, B., Is Newton's second law really Newton's? American Journal of Physics, **79,** 10–15, 2011.
3. Pourciau, B., Newton's interpretation of Newton's second law, Arch. Hist. Exact Sci. **60**, 157–207 (2006).
4. Sharma, A., Acta Ciencia Indica, Vol. XXV P, No. 4, 181 (1999).

# 9

# Experimental confirmations of equations of conservation laws in elastic collisions

## First glimpse

- ✓ The equations based on elastic collisions take into account directly mass, but no other characteristics of bodies and experimental set up.
- ✓ The important thing is to experimentally confirm equations.
- ✓ If a projectile is very-2 heavy than the target, then after collision target must move with double velocity.
- ✓ When two bodies of equal masses collide, then they exchange their velocities.
- ✓ If the equations of the conservation of momentum and kinetic energy are written but not experimentally confirmed, then the job is half done.
- ✓ It is stressed that at the macroscopic level the conditions under which these equations hold good must be determined.
- ✓ It can be experimentally checked up to which extent the composition of bodies and the resistive forces of the system play a significant role in collisions.
- ✓ The study of elastic collisions (conservation of momentum and kinetic energy) is incomplete if equations are not experimentally verified.
- ✓ Mathematical equations based on elastic collisions predict that the velocity of the projectile and the target after collisions is uniform, i.e. acceleration is zero.

## 1.0. Laws of conservation in nature

The conservation laws are not only the basis of the scientific formulations but also very essential for the stability of a system or society. For example, let us imagine the state of affairs in the most ancient society. Let one person *A* snatched a fruit from the other person *B*. This high handedness, disliked by *B*, compelled him to take the fruit back from *A* and both may even hurt each other in the process. In the process, *B*, as far as possible, would have taken back the fruit from *A* or the equivalent amount of other eatables, such as flowers or sweet roots, etc. Therefore, whatever are the initial contents with B before the tussle, as far as possible must be equal to final contents after the tussle; it is an essential requirement for the peace and contentment between two. The same is the principle of the judiciary system globally. This phenomenon can be understood or discussed in view of the general law of conservation. According to this perception

$$\text{Initial quantity before event} = \text{Final quantity after event} \quad (9.1)$$

This is an example from ancient times, as the civilization grew, some consensus or understanding was established that 'this' quantity is equivalent to 'that' quantity. For example, an apple is equivalent to two mangoes or four guavas. In addition, if the person works in the fields then his daily labour is equivalent to a few kilograms of grain. Thus, the concept of equivalence of various quantities is as old as the civilization itself. In addition, it will continue as long the civilization lasts hopefully forever in one form of other. Nowadays one common quantity that can be exchanged in terms of other quantity is known as currency, which is globally recognized, say the dollar or Euro or new Indian rupee, etc.

In general, the law of conservation may be expressed in basic physics in terms of the principle of dimensional homogeneity. This principle of dimensional homogeneity implies that dimensions in each term of the LHS and RHS of the equation are the same. If the dimensions in each term of LHS and RHS are not the same, then the equation is not regarded as correct. In addition, in chemical reactions, if the numbers of atoms in the reactants and products are not equal then the reactions are not balanced, it is the application of conservation laws in terms of atoms.

The similar resemblance exists in mathematical equations. In an equation the left hand side (LHS) is always equal to the right hand side (RHS) or may be understood as the law of mathematical conservation. These are a few elementary or qualitative or approximate examples of conservation laws just for understanding that such ideas existed since the inception of humanity or civilization in different ways. Nowadays, these are expressed scientifically in terms of more specific and quantitative quantities in each scientific discipline.

Here we will discuss the conservation of momentum and kinetic energy, both theoretically and experimentally.

### *1.1. Linear momentum*

The linear momentum is defined as the total quantity of motion contained in a body and is measured as the product of mass of body and velocity.

If a body of mass $m$ is moving with velocity $v$ then its momentum, $P$ is given by

$$P = mv \tag{9.2}$$

The second law of motion is completely dependent on the momentum.

Let $P$ be the momentum of the body and $F$ is the external force acting on it, then according to the second law of motion

$$F = \frac{d(mv)}{dt} = \mathrm{m}\frac{dv}{dt} \tag{9.3}$$

or

$$F = ma \tag{9.3}$$

where $a$ is acceleration, i.e. the rate of the change of velocity. The unit of force in the SI system is Newton (in honour of Sir Isaac Newton) and in the CGS system it is dyne.

### *1.2. Principle of conservation of linear momentum*

The law of conservation of the linear momentum states that if no external forces act on the system then the momentum of the system remains constant.

It is understood on the basis of the second law of motion. According to this law, the rate of the change of momentum is the force and is given by

$$F = \frac{dv}{dt} \tag{9.3}$$

In the absence of external force, $F = 0$

$$0 = \frac{dv}{dt}$$

$$0 = \text{rate of change of momentum}$$

As the change in momentum $p$ is zero hence it is constant

$$\text{momentum } (mv) = \text{constant}$$

$$p = mv = \text{constant} \tag{9.4}$$

Momentum of the system in initial stage
= Momentum of system at final stage

Hence the linear momentum is conserved when no external forces act on the body.

**Example:** When a bullet is fired from a gun, the gun recoils or kicks back in the backward direction. The reason can be given on the basis of conservation of the linear momentum

**Explanation:** Let $M$ be the mass of the gun and $m$ the mass of bullet. Initially, both the gun and the bullet are at rest, hence both velocity and momentum are zero. Let us calculate the initial and final momentum of the system.

$$\begin{aligned}\text{Initial momentum} &= \text{momentum of gun} + \text{momentum of bullet}\\ &= M.0 + m.0\\ &= 0 + 0\end{aligned}$$

Let after firing the bullet moves with velocity $v$ and the gun with velocity $V$.

$$\begin{aligned}\text{Final momentum} &= \text{momentum of the gun} + \text{momentum bullet}\\ &= MV + mv\end{aligned} \tag{9.5}$$

According to the law of conservation of the momentum

Momentum of the system in initial stage
= Momentum of system at final stage

$$0 = MV + mv$$

$$V = -\frac{mv}{M} \tag{9.6}$$

The negative sign indicates that as the bullet moves in the forward direction the gun moves in the backward direction. Hence, the directions of the gun and the bullet are opposite.

**Numerical example:** Let a shell of mass 0.01 kg is fired from a gun with a mass of 25 kg. If the muzzle velocity of the shell is 200 $ms^{-1}$, what is the recoil speed of the gun?

**Solution:** Mass of the shell, $M_s = 0.01$ kg
Mass of the gun, $M_g = 25$ kg

Initially, both the bullet and the gun are at rest hence,

Initial momentum = momentum of shell + momentum of gun
= 0.01 × 0 + 25 × 0
= 0 + 0

After firing the shell moves with velocity 200 m/s and the gun recoils with velocity $V$ (say),

Final momentum
= Final momentum of bullet + final momentum of gun
= 0.01 kg × 200 + 25 × $V$

The conservation of the linear momentum requires

Initial momentum of the system
= Final momentum of the system

$$V = -\frac{0.01 \times 200}{25} = -\,0.08 \text{ m/s or } -8 \text{ cm/s} \quad (9.7)$$

The negative sign indicates that the direction of the gun is the opposite to that of the shell, i.e. the shell moves forward and the gun moves backward. Unless the above prediction is not quantitatively confirmed, it should not be taken as granted that actually the body recoils with a velocity of 8 cm/s.

## *1.3. Questions to be pondered upon*

The numerical calculations like those above have been done manytimes in physics and will be done numerous times in future also. But the question of interest is that whether actually the gun recoils with a velocity of 8 cm/s or not. But if the gun recoils with the predicted velocity, then under which conditions it does so? What is the role of resistive forces in this regard? It is the most significant aspect of the observation or demonstration, but has not been mentioned at all in the existing literature. Thus, here the system of the shell and the gun is supposed to be placed over the ideal surface. But what are characteristics of the ideal system at which the body recoils with velocity $v$, 8 cm/s. Now placing the gun at various surfaces the recoil velocity can be calculated.

(i) When the gun is placed at the rough ground then firing the gun the recoil velocity is calculated. Let it be denoted by $V$ (ground).

(ii) Further, the gun is placed in a boat or on a wooden plank in water in a lake and fired. When the bullet is fired it moves forward and, to conserve the momentum, the gun recoils. The recoil velocity of the gun is then calculated. Anyhow, this case is technically more complex compared with other measurements. Let the recoil velocity be denoted by *V* (water).

(iii) The gun is then placed on a smooth wooden floor and fired. When the bullet is fired it moves forward and, to conserve the momentum, the gun recoils. The recoil velocity of the gun is calculated. Let it be denoted by *V* (wood).

(iv) Further, the gun is placed on a glass floor or surface and fired. When the bullet is fired it bullet moves forward and, to conserve the momentum, the gun recoils. The recoil velocity of the gun is calculated. Let it be denoted by *V* (glass).

(iv) The gun is then placed on an ice surface and fired. When the bullet is fired it bullet moves forward and, to conserve the momentum, the gun recoils. The recoil velocity of the gun is calculated. Let it be denoted by *V* (ice).

Other surfaces can also be chosen for experimentation and to draw wider conclusions. In the numerical calculations as in Eq. (9.7), nothing is mentioned at all about the surface on which the gun is placed. The conceptual basis is independent of the nature of the surface. Also it is not mentioned whether the gun is fitted with wheels or pushes itself backward. So Eq. (9.7) or conservation of momentum is theoretically valid for every surface, may be rough or smooth. However, practically it is not feasible. The nature of the surface of the medium and resistive forces of the system is the most significant aspect which is not addressed. To understand the concept fully such basic experiments has to be performed at various surfaces, with the most sensitive equipment.

As in the mathematical calculations nothing is mentioned about the surface on which the body is placed or resistive forces of the system, thus all the velocities must be equal

$$V\text{ (ground)} = V\text{ (water)} = V\text{ (wood)} = V\text{ (glass)} = V\text{ (glass)} = 8\text{ cm/sec}$$

or

$$V\text{ (typical surface)} = 8\text{ cm/sec} \tag{9.8}$$

But it is a simple conclusion that the recoil velocity practically will be less for the typically rough surface than the smooth glass surface.

Thus magnitude of the recoil velocity may or may not be satisfied. It is theoretical analysis only, however experiments are required to be conducted to find out the surface for which predictions are precisely justified. It will help in quantitative understanding of the phenomena. The conservation of momentum is the basic law of physics and the well taught law but the role of the surface (where the body is placed) is not discussed at all.

## 1.4. Work

French mathematician Gustave-Gaspard Coriolis [1792–1843] was first to define work as the product of force and displacement:

$$W = FS\cos\Phi \tag{9.9}$$

where $W$ is work, $\Phi$ is the angle between force and displacement $S$. The equations of various forms of energy are based on Eq. (9.9). If the direction of force and displacement is same, i.e. $\Phi = 0$, then the work done is maximum, i.e.

$$W = FS\cos 0 = FS.\ 1 = FS = maS \tag{9.10}$$

This is the equation given by Coriolis which predicts that when a coolie takes a head load then does no work (the component of force along the direction of displacement is 90°, cos 90° = 0), i.e.

$$W = FS\cos\Phi = FS\cos 90° = 0.$$

Here the definition, units and dimensions of work are defined in the simplest possible way. Work is said to be done by a force when the body is actually displaced through some distance in the direction of applied force.

### The simplest case

In the simplest form the work is written as

$$W = FS = MaS \tag{9.10}$$

where $W$ is work, $S$ is the distance travelled, $F$ is force, $m$ is the mass of the body and $a$ is acceleration.

(i) If a person tries to push the wall but is unable to move it, then he does no work. The wall does not move, i.e. $S = 0$. Hence,

$$W = FS = F.\ 0 = 0$$

(ii) Let a person pushes the car in which four passengers were sitting into a river, then he does work. In terms of physics he does useful work, as $F$ and $S$ both are non-zero.

(iii) A coolie taking an 80 kg suitcase on his head travels through a distance of 1 km does no work.
In terms of the mathematical equation of physics the coolie does no work. In this case the distance travelled is non-zero but the 'component of the force' in the direction of distance is zero, i.e. $FS \cos 90^\circ = 0$. Thus,

$$W = FS = 0.S = 0$$

Hence the result.

(iv) The second part of Newton's first law of motion states that the body continues in the state of uniform motion; under this condition no work is done. It is justified below. The work done can be written in the form of acceleration

$$a = \frac{(v-u)}{t}$$

where $u$ is initial velocity, $v$ is terminal velocity of the body and $t$ is time. When $u = v$, $a = 0$, $F = 0$. Thus

$$W = FS = M\frac{(v-u)}{t}S = 0$$

If the body moves with uniform velocity (the second part of the first law of motion states the body maintains uniform velocity, if the external force acts on it) then

$$v = u \tag{9.11}$$

or

$$W = FS = M\frac{(v-u)}{t}S = 0 \tag{9.12}$$

So even if the body is displaced in the direction of force and maintains uniform velocity no work is done. This aspect is not addressed in the existing physics. The reason for this result is the perception that the body maintains uniform velocity ($u = v$, $a = 0$) without force if once set in motion. Thus, the conditions of applicability of Newton's first law of motion (body persists with uniform velocity) means the work done is zero. The law can also be defined in terms of zero work done, as above.

### *1.5. Units and dimensions of work*

The simplest way of defining work is,

$$\text{Work} = \text{Force} \times \text{Distance} \tag{9.9}$$

Further, the dimensional equation can be written as

$$\text{Dimensional equation, work} = MLT^{-2} \times L = ML^2T^{-2}$$

(i) In the CGS system (absolute unit) the unit of work is erg. The work is said to be one erg when a force of one dyne actually moves the body through a distance of 1 cm in the direction of applied force

$$1 \text{ erg} = 1 \text{ dyne} \times 1\text{cm} = \text{dyne cm}$$

One dyne is that much force which produces an acceleration of 1 cm/s$^2$ in a body of mass 1 gm.

$$1 \text{ erg} = 1 \text{ dyne} \times 1 \text{ cm} = \text{dyne cm} = 1 \text{ gm cm}^2 \text{ s}^{-2}$$

(ii) In the SI system (absolute units) the unit of work is Joule, $W = FS$
The work is said to be one Joule when a force of one newton actually moves the body through a distance of one meter in the direction of applied force,

$$1 \text{ Joule} = F.S = 1 \text{ newton} \times 1 \text{ metre} = \text{N m}$$
$$1 \text{ Joule} = 10^5 \text{ dyne} \times 10^2 \text{ cm} = 10^7 \text{ dyne cm}$$

### *1.6. Kinetic energy*

Kinetic energy is the energy possessed by the body by virtue of its motion. It is another term which is useful in understanding the conservation laws in elastic collisions. Examples of kinetic energy are:

1. The kinetic energy of air is used to run windmills
2. The kinetic energy of moving a stone is used to break the targets
3. The kinetic energy of a hammer is used in driving a nail into a piece of wood
4. A bullet from a gun can pierce the target due to kinetic energy

**Measurements of kinetic energy**

Consider a body of mass $M$ is lying at rest. Let an external force acts on the body such that its velocity $v$ is maintained in the direction of force. Let the body travels distance $S$ in time $t$. Then the work done by the force in displacing the body of mass $M$ though distance $S$,

$$W = FS \tag{9.9}$$

Let $a$ be the acceleration produced in the body, then according to Newton's second law of motion,

$$F = Ma \tag{9.3}$$

Substituting Eq. (9.3) in Eq. (9.9)

$$W = MaS$$

The value of $a$ can be calculated from the relation (meant for the uniformly accelerated body)

$$v^2 - u^2 = 2aS$$

or
$$v^2 - 0 = 2aS$$

or
$$a = \frac{v^2}{2S} \tag{9.13}$$

$$W = MaS = M \times \frac{v^2}{2S}\, S = \frac{1}{2} Mv^2 \tag{9.14}$$

This work is stored in the body of mass $M$ in the form of kinetic energy and serves as the expression for kinetic energy.

Units and dimensions of kinetic energy

Dimensional equation, $KE = \frac{1}{2} MV^2 = M(LT^{-1})^2 = ML^2T^{-2}$

These are the same as the dimensions of work

The energy of the body is the capacity to do work. It has the units and dimensions of work. The principle of conservation of energy implies that: "it states that the energy can neither be created nor be destroyed but can be changed from one form to other."

## 2.0. Newton's first law of motion or the law of conservation of velocity

Newton's first law of motion is stated in original Latin:

**Lex I:** *Corpus omne perseverare in statu suo quiescendi vel movendi uniformiter in directum, nisi quatenus a viribus impressis cogitur statum illum mutare.*

Newton's first law of motion as translated in English:

'Every body persists in its state of being at rest or of moving uniformly straight forward, except insofar as it is compelled to change its state by force impressed.'

The second part of Newton's first law of motion implies that: 'every body persists in its state of moving uniformly straight forward, except insofar as it is compelled to change its state by force impressed.'

or

'Every body persists in its state of moving uniformly straight forward, if external force, $F = 0$.'

This deduction is in the absence of external force ($F = 0$). Thus

$$v = \text{uniform} \quad \text{or} \quad u = v \tag{9.11}$$

$$\text{Initial velocity} = \text{Terminal velocity}$$

i.e. in the absence of the external force the body moves with uniform velocity. So it may also be termed as the law of conservation of velocity. It can be easily justified that the law of conservation of momentum and kinetic energy also follows from the first law of motion.

## *2.1. Newton's first law of motion, law of conservation of momentum, and law of conservation of kinetic energy*

If the concepts of the first law of motion, conservation of momentum and the law of conservation of energy are deeply analyzed it is seen that all the three are basically the same.

(i) It is evident that the first law of motion implies that the body maintains its state of uniform motion. But it is silent about the fact how the body is set in motion, initially in uniform motion. Definitely the external force is required to set the body in motion. Newton's first law of motion states that if the body is in uniform motion, then this uniform motion is preserved if no external force acts on it.

$$v = \text{uniform}$$

$$\text{or} \qquad u = v \tag{9.1}$$

(ii) Also the conservation of momentum states that as in Eq. (9.4) in the absence of external force ($F = 0$)

$$p = \text{constant}$$

$$\text{Initial momentum} = \text{Final momentum}$$

Multiplying both sides of Eq. (9.11) by mass $m$

$$mu = mv$$

$$\text{Momentum of body} = \text{constant} \tag{9.4}$$

or

$$u = v$$

which is nothing else but Eq. (9.1).

Thus law of conservation of velocity follows from the first law of motion.

(iii) The law of conservation of kinetic energy, i.e. energy, remains constant

$$\text{Kinetic energy} = \text{Constant}$$

$$\text{Initial } KE = \text{Final } KE$$

Also it follows from Eq. (9.11), i.e. squaring it

$$u^2 = v^2 \tag{9.15}$$

Multiplying both sides of Eq. (9.15) with $\frac{1}{2}M$

or

$$\frac{1}{2}Mv^2 = \frac{1}{2}Mu^2$$

Initial kinetic energy = Final kinetic energy

$$\text{Kinetic energy} = \text{constant} \tag{9.16}$$

or

$$v^2 = u^2$$

$$v = u \tag{9.1}$$

which is nothing else but the law of conservation of velocity.

Thus, all the three cases, i.e. the first law of motion, conservation of momentum and conservation kinetic energy are the same. In all above cases the resistive forces play a significant role. In the first law of motion it is assumed that the body automatically remains in motion (the body continues in the state of uniform motion, in the absence of external force) if once set in motion. It must be noted that body preserves its state of uniform motion (if no external force acts on it); but is silent about how the body is initially set in uniform motion. Obviously the force at the outset is required to set the body in motion.

## 3.0. Collisions

The collision between two bodies is defined as the mutual interaction of the bodies for the interval of time as result of which the energy and momentum of interacting bodies change

Collision between two billiard balls or between two automobiles on the road are few examples of collisions in daily life. In case of Rutherford's alpha scattering experiments, the alpha particles are scattered due to electrostatic interaction between the alpha particle and nucleus from the distance. In this collision there is no physical contact between the alpha particle and the nucleus. Also similar poles of a magnet repel each other without physical contact; this is also an example of collision.

Thus, in physics, a collision is said to have occurred if the bodies physically collide against each other or even when the path of motion of one body is affected by the other. There are three types of collisions. But here discussion is limited to the one-dimensional elastic collisions only. The theoretical predictions of one-dimensional collisions must be experimentally justified. The one-dimensional elastic collisions are most suitable for the purpose because the experimental study of one-dimensional elastic collision is extremely important and simple in understanding the conservation laws.

### *3.1. Elastic collision*

The collisions in which both the momentum and kinetic energy of the system are conserved, are called elastic collisions.

Characteristics of elastic collisions:

(i) The momentum is conserved

(ii) The kinetic energy is conserved

(iii) The total energy is conserved

(iv) The coefficient of restitution or coefficient of resilience is unity

$$e = \frac{v_2 - v_1}{u_1 - u_2} \tag{9.17}$$

where $u_1$ is the initial velocity of the projectile, $u_2$ is the initial velocity of the target, $v_1$ is the terminal velocity of the projectile, $v_2$ is the terminal velocity of target.

Collisions between hard steel balls are nearly elastic. In a perfectly elastic collision the coefficient of restitution is regarded as unity.

### 3.2. Inelastic collisions

The collisions, in which the momentum of the system is conserved but kinetic energy is not conserved, are called inelastic collisions.

The majority of collisions in daily life are inelastic collisions. The kinetic energy appears to be lost in inelastic collisions appear in some other forms of energy, such as heat and sound energy, etc.

### 3.3. Perfectly inelastic collisions

The collisions, in which colliding bodies stick with each other after collision and move with common velocity, are called perfectly inelastic collisions.

Mud thrown on the wall and sticking to it, a bullet fired into a wooden block and remaining embedded in it, are examples of perfectly inelastic collisions. It is believed in such cases the linear momentum is conserved but loss of kinetic energy is maximum.

### 3.4. Coefficient of restitution or coefficient of resilience

In practice, collisions between all real objects are neither perfectly elastic nor they are perfectly inelastic. They are called imperfect or semi-elastic collisions. The degree of elasticity of collisions is determined by a quantity called the coefficient of restitution or the coefficient of resilience: 'it is defined as the ratio of relative velocity of separation after collision to the relative velocity of approach. It is represented by '$e$'

$e$ = Relative velocity of separation (after collision)/
relative velocity of approach (before collision)

$$e = \frac{v_2 - v_1}{u_1 - u_2} \tag{9.17}$$

where $u_1$ and $u_2$ are velocities of two bodies before collision; $v_1$ and $v_2$ are respective velocities after collision.

(i) For a perfectly elastic collision, the relative velocity of separation after the collision is equal to the relative velocity of approach before the collision.

$$e = 1 \tag{9.18}$$

(ii) For a perfectly inelastic collision, the relative velocity of separation after the collision = 0

$$e = 0 \tag{9.19}$$

(iii) If $u_1 = u_2$, i.e. body $A$ (projectile) and body $B$ (target) move with the same velocity then the collision does not occur. Hence the coefficient of restitution is not defined.

### *3.5. The aim of forthcoming discussion*

For the sake of simplicity here we will confine our discussion to one-dimensional elastic collisions only. In such collisions the linear momentum and kinetic energy are both conserved. In addition, the coefficient of restitution or coefficient of resilience $e$, as given by Eq. (9.17), is also unity. Thus, this discussion is related with fundamental laws.

Theoretically, it is easier to discuss and outline the conditions under which the conservation laws, i.e. Newton's first law of motion (conservation of velocities), conservation of linear momentum and conservation of kinetic energy hold good. But the actual aim is to experimentally confirm the predictions of equations in elastic collisions.

(i) Ideally, if the predictions of equations based on elastic collisions are experimentally confirmed, then the conservation laws are obeyed.

(ii) If the predictions of equations based on elastic collisions are not experimentally obeyed, then sources of errors may exist. If the difference between predictions and contribution due to the sources of error is the same, then the results are consistent within the realm of errors.

(iii) If the predictions of equations based on elastic collisions are not experimentally obeyed, then sources of errors may exist. If the difference between predictions and contribution due to the sources of error is not the same, then the results are not consistent. The contributions of the sources of errors must be precisely estimated to draw final conclusions.

Apparently, in the existing literature there is no experimental data which may address the above queries quantitatively. Thus, it is stressed that specific and quantitative experiments should be conducted. It is a simple scientific principle that mathematical predictions have to be experimentally confirmed in all the equations.

### *3.6. One-dimensional elastic collision: mathematical equations*

The collision in which both the momentum and kinetic energy are conserved and the colliding bodies continue to move in the same straight line after collision is the one-dimensional elastic collision.

Consider two perfectly elastic bodies $A$ and $B$ of masses $M_1$ and $M_2$ moving along the same straight line with velocities $u_1$ and $u_2$ respectively. The bodies will collide only if $u_1 > u_2$. The composition of bodies or characteristics of media are immaterial as far the equations are concerned.

The two bodies undergo a head-on collision and continue moving along same straight line with velocities $v_1$ and $v_2$ in the same direction. Two bodies will separate after collisions only if $v_2 > v_1$.

As in elastic collisions the momentum is conserved, we have

$$M_1u_1 + M_2u_2 = M_1v_1 + M_2v_2 \tag{9.20}$$

Since the kinetic energy is also conserved, we have

$$\frac{M_1u_1^2}{2} + \frac{M_2u_2^2}{2} = \frac{M_1v_1^2}{2} + \frac{M_2v_2^2}{2} \tag{9.21}$$

From Eq. (9.20) we have

$$M_1(u_1 - v_1) = M_2(v_2 - u_2) \tag{9.22}$$

Also from Eq. (9.21) we can write

$$M_1(u_1^2 - v_1^2) = M_2\,(v_2^2 - u_2^2) \tag{9.23}$$

Dividing Eq. (9.23) by Eq. (9.22)

$$\frac{\left(u^2{}_1 - v_1^2\right)}{\left(u_1 - v_1\right)} = \frac{\left(v_2^2 - u_2^2\right)}{\left(v_2 - u_2\right)}$$

$$u_1 + v_1 = v_2 + u_2$$

$$u_1 - u_2 = v_2 - v_1 \tag{9.24}$$

From the equation it follows that in one-dimensional elastic collision, *the relative velocity of approach* $u_1 - u_2$ before collision is equal to the *relative velocity of separation* $v_2 - v_1$ after collision.

### *3.7. Coefficient of restitution or coefficient of resilience*

The coefficient of restitution can be determined by dividing both sides of Eq. (9.24) with $u_1 - u_2$ as

$$e = \frac{v_2 - v_1}{u_1 - u_2} \tag{9.17}$$

For elastic collisions by definition we have,

$$e = \frac{v_2 - v_1}{u_1 - u_2} = 1 \tag{9.18}$$

### *3.8. Final velocities of bodies* $\mathbf{v}_1$ *and* $\mathbf{v}_2$

Let us find the velocity of body *A* after collision. From Eq. (9.24) we have

$$v_2 = u_1 - u_2 + u_2$$

Substituting $v_2$ in Eq. (9.20), we get

$$\begin{aligned} M_1u_1 + M_2u_2 &= M_1v_1 + M_2(u_1 - u_2 + v_1) \\ &= M_1v_1 + M_2u_1 - M_2u_2 + M_2v_1 \\ (M_1 - M_2)u_1 + 2M_2u_2 &= (M_1 + M_2)v_1 \end{aligned}$$

or
$$v_1 = \frac{(M_1 - M_2)u_1 + 2M_2u_2}{(M_1 + M_2)} \tag{9.25}$$

Again from Eq. (9.24)

$$v_1 = v_2 - u_1 + u_2 \tag{9.26}$$

Substituting $v_1$ from Eq. (9.26) in Eq. (9.20)

$$\begin{aligned} M_1u_1 + M_2u_2 &= M_1(v_2 - u_1 + u_2) + M_2v_2 \\ &= M_1v_2 - M_1v_1 + M_1u_2 + M_2v_2 \\ (M_2 - M_1)u_2 + 2M_1u_1 &= (M_1 + M_2)v_2 \end{aligned}$$

$$v_2 = \frac{(M_2 - M_1)u_2 + 2M_1u_1}{M_1 + M_2} \tag{9.27}$$

The known value of initial velocities can be imparted to body *A* and body *B* taking into account the values of $v_1$, $v_2$, $M_1$ and $M_2$

## 4.0. The values of $v_1$ and $v_2$ under special conditions

We have already considered two perfectly elastic bodies *A* and *B* of masses $M_1$ and $M_2$ moving along the same straight line with velocities

$u_1$ and $u_2$ respectively. The bodies will collide only if $u_1 > u_2$. The two bodies undergo a head-on collision and continue moving along the same straight line with velocities $v_1$ and $v_2$ in the same direction. Two bodies will separate after collisions only if $v_2 > v_1$. Let us calculate the final velocities of the two bodies ($v_1$, $v_2$) after collision in the following special cases. The special cases will depend upon the limiting values of $u_i$'s and $M_i$'s. The general equations for the final velocities of body *A* and body *B* are given by Eq. (9.25) and Eq. (9.27)

## 4.1 When two bodies are of equal masses

Let us consider the case when the masses of body *A* (projectile) and body *B* (target) are equal, i.e. $M_1 = M_2 = M$ (say)

From Eq. (9.25), we have

$$v_1 = \frac{(M - M\ )u_1 + 2Mu_2}{(M + M\ )} = \frac{0 + 2Mu_2}{M + M}$$

$$v_1 = u_2 \tag{9.28}$$

The terminal velocity of the projectile (body *A*) = Initial velocity of the target (body *B*).

Also from Eq. (9.27), we have

$$v_2 \frac{(M - M)u_2 + 2Mu_1}{M + M} = \frac{0 + 2Mu_1}{2M}$$

$$v_2 = u_1 \tag{9.29}$$

The terminal velocity of the target (body *B*) = Initial velocity of the projectile (body *A*).

Thus, if two bodies of equal masses suffer one-dimensional elastic collision, after the collision they will exchange the velocities.

Practically, it implies that two buses *A* and *B* of equal masses (160 kg, say) are moving with a velocity of 15 km/h and 5 km/hr and collide on the road. Then after the collision bus *B* will move with a velocity of 15 km/h and bus *A* with a velocity of 5 km/h.

### *4.2. Masses of bodies are different and body B (target) is at rest, $u_2 = 0$*

Let the body *A* of mass $M_1$ is moving with velocity $u_1$, and body *B* of mass $M_2$ is at rest, then what would the values of $v_1$ and $v_2$ be? The values of $v_1$ and $v_2$ can be calculated from Eq. (9.25) and Eq. (9.27) setting $u_2 = 0$,

$$v_1 = \frac{(M_1 - M_2)u_1}{(M_1 + M_2)} \tag{9.30}$$

$$v_2 = \frac{2M_1u_1}{M_1 + M_2} \tag{9.31}$$

When the target body *B* is at rest, the final velocities of the two bodies can be calculated in two cases depending upon the values of $M_1$ and $M_2$.

### 4.2 (a). Masses of bodies are same and body B (target) is at rest, $u_2 = 0$

Let the body *A* is moving with velocity $u_1$, and body *B* is at rest ($u_2 = 0$), such that the masses of two bodies are the same, i.e. $M_1 = M_2 = M$. Then what would the values of $v_1$ and $v_2$ be? These can be calculated by setting $M_1 = M_2 = M$ in Eq. (9.30) and Eq. (9.31)

$$v_1 = \frac{(M - M)u_1}{(M + M)} = 0 \tag{9.32}$$

$$v_2 = \frac{2Mu_1}{M + M} = u_1 \tag{9.33}$$

Therefore, we find that when the body *A* collides against the body *B* of equal mass at rest, then the body *A* should come to rest and the body *B* should move on with the velocity of the body *A*.

Consider two buses *A* and *B* of equal masses (160 kg, say). The bus *A* is moving with velocity 15 km/h and the bus *B* is at rest ($u_2 = 0$). Both buses collide, then bus *A* comes to rest ($v_1 = 0$) and bus *B* moves with velocity of bus *A* ($u_1$), i.e. 15 km/s.

### 4.2(b). When mass of body B is negligible compared to that of body A ($M_2 << M_1$)

Let the body *A* is moving with velocity $u_1$, and body *B* is at rest ($u_2 = 0$), such that the mass of body *B* is negligible compared to the mass of body *A*, i.e. $M_2 << M_1$. What would be the values of $v_1$ and $v_2$?

When $M_2 << M_1$ (the target is small), then in Eq. (9.30) and Eq. (9.31), $M_2$ can be neglected compared to $M_1$, i.e.

$M_1 - M_2 = M_1$, $M_1 + M_2 = M_1$. Therefore, Eq. (9.30) and Eq. (9.31) become

$$v_1 = \frac{M_1 u_1}{M_1} = u_1 \tag{9.34}$$

$$v_2 = \frac{2M_1 u_1}{M_1} = 2u_1 \tag{9.35}$$

Therefore, when the body *A* collides against the light body *B* at rest (very-2 small mass comparatively), the body *A* keeps on moving with the same velocity and the body *B* should start moving with the velocity double that of body *A*.

Consider a train of mass of 2000 kg (say) and a ball of mass 50 gm. The train is moving with a velocity of 150 km/h and the ball is hanging (with a cotton thread) in the track of the train. As the train collides with the ball (at rest, $u_2 = 0$), then the ball must move with a velocity of 300 km/h and the train with the original velocity of 150 km/h.

The terminal velocity of the ball can be calculated by estimating the force with which the ball is hit. Or the force can be calculated from the magnitude of acceleration ($u_2 = 0$, $v_2 = 300$ km/h) produced in the ball. However, Eq. (9.35) predicts that when the projectile strikes the target, then the target moves with uniform velocity $2u_1$.

## 5.0. Experimental confirmation of the equation based upon conservation laws

Any theory or equation gets complete scientific validity after it is experimentally verified in repeated experiments or it is pointed under which conditions it holds good without authenticity. In most cases equations are formulated under some conditions. If equations are written *n* times it does not mean they are quantitatively correct and thus experiments must be conducted to counter check the predictions. Mathematically, the above equations are applicable to all bodies of different composition, may be of cloth, steel, plastic, etc. on various surfaces. In the experiments all possible factors influencing the results have to be taken into account. The main factors which can affect results can be short listed as

(i) The roughness of the surfaces on which experiments are conducted. Some surfaces have resistive forces, which try to destroy the motion over its surface. If the value of resistive force is maximum, then the velocity of the body is the least. It is established fact. But the predictions based upon the mathematical equations do not at all take the roughness of the surface into account or do not comment on it at all. But without specific description of resistive forces the conceptual study is incomplete.

(ii) The mathematical equations only take into account the masses and velocities of bodies, not the shapes, such as square, triangular, spherical or semi-spherical or irregular etc.

(iii) We can have numerous possibilities or combinations for compositions of nature of bodies.

(a) Body *B* can be of cloth and body *A* of wood
(b) Body *B* can be of wood and body *A* of wood
(c) Body *B* can be of plastic and body *A* of wood
(d) Body *B* can be air-filled football and body *A* of wood
(d) Body *B* can be of water-filled ball and body of wood
(e) Body *B* can be steel and body *A* of the wood
(f) Or different combinations of the bodies can be considered on the different surfaces which have different resistive forces.

Consider a soft ball of mass 5 gm moving with a velocity of 5 cm/s hit by a steel ball of mass 20 gm. The values of $v_1$ and $v_2$ are noticed or measured. After this let there is another collision by a body of steel of mass 5 gm moving with a velocity of 5 cm/s with a 20 gm steel target. Now the velocities $v_1$ and $v_2$ are different. Hence, the nature of composition of bodies is very significant in addition to masses and velocities.

Such experiments can be conducted in numerous ways. The equations must be experimentally confirmed so that concrete conclusions about the applicability of laws of the conservation of momentum and kinetic energy can be drawn. The mathematical treatment is completely silent about atmospheric disturbances; the kinetic energy of wind can definitely affect the results. Also, the roughness of the surfaces in which bodies move is not specified at all.

(vi) In a collision it is also important to notice where the projectile (body *A*) hits the target (body *B*). Ideally, the projectile must hit the target at the centre, only then the collision is one-dimensional. Consequently, the role and importance of the point of impact can be determined. Needless to mention, all the movements are considered in one dimension only. For experimental verification of equations many factors have to be precisely taken into account. It is absolutely necessary to conduct such experiments to understand the basic laws of physics practically and experimentally. Mathematically, the coefficient of restitution is unity in all such cases.

### *5.1. The general values of u and v*

Consider a body *A* (projectile) of mass 4 gm is moving with a velocity of 10 cm/s, and body *B* (target) of mass 8 gm with a velocity of 4 cm/s. Thus the given values of parameters are

$$M_1 = 4 \text{ gm}, M_2 = 8 \text{ gm},$$
$$u_1 = 10 \text{ cm/s}, u_2 = 4 \text{ cm/s}$$

Here the aim is to calculate the various values of velocities, i.e. after collision the velocity of body *A* is $v_1$ and that of body *B* is $v_2$, for different values of various parameters.

We have Eq. (9.25) which gives the value of the terminal velocity of the projectile (body *A*) and the terminal velocity of the target (body *B*) is given by Eq. (9.27).

$$v_1 = \frac{(M_1 - M_2)u_1 + 2M_2u_2}{(M_1 + M_2)} \tag{9.25}$$

and

$$v_2 = \frac{(M_2 - M_1)u_2 + 2M_1u_1}{M_1 + M_2} \tag{9.27}$$

We have to calculate the followings from equations based upon collisions, i.e. Eq. (9.25) and Eq. (9.27)

$$v_1 = ?$$
$$v_2 = ?$$

Now value of $v_1$ can be calculated from Eq. (9.25)

$$v_1 = \frac{(M_1 - M_2)u_1 + 2M_2u_2}{(M_1 + M_2)} = \frac{24}{12} = 2 \text{ cm/s} \tag{9.36}$$

Further the value of $v_2$ is given by Eq. (9.27)

$$v_2 = \frac{(M_2 - M_1)u_2 + 2M_1u_1}{M_1 + M_2} = \frac{96}{12} = 8 \text{ cm/s} \tag{9.37}$$

Now the next question is to see under which conditions the values of $v_1$ and $v_2$ are 2 cm/s and 8 cm/s respectively. If the bodies move with the above velocities at all surfaces, then the conservation of linear momentum and kinetic energy holds good for all cases. If it is not so then the conditions under which the conservation laws hold good have to be investigated. It may require repeated and sensitive experiments. As the collision is elastic, so the coefficient of restitution is unity. Thus the value of *e* can be calculated from Eq. (9.17)

$$\text{Coefficient of restitution, } e = \frac{v_2 - v_1}{u_1 - u_2} = \frac{(8-2)\text{ cm/s}}{(10-4)\text{ cm/s}} = 1$$

### *5.1(a). When two bodies are of equal masses, $M_1 = M_2 = M$*

Now consider the case when the masses of both bodies, i.e. body *A* (projectile) and body *B* (target), are equal such that $M_1 = M_2 = M$ (say) As in the previous case, the body *A* is moving with a velocity of 10 cm/s ($u_1$) and body *B* with a velocity of 4 cm/s ($u_2$).

The terminal velocity of body *A*, i.e. $v_1$, can be calculated from Eq. (9.25) by substituting various values.

$$v_1 = \frac{(M - M)u_2 + 2M\,u_2}{(M + M)} = \frac{0 + 2Mu_2}{2M}$$

$$v_1 = u_2 = 4\text{cm/s} \tag{9.38}$$

Terminal velocity of body *A* (projectile) = Initial velocity of body *B* (target) = 4 cm/s

Let body *A* and body *B* are so fabricated that their masses are equal. Thus, the terminal velocity of body *A* ($v_1$) is precisely equal to 4 cm/s. It can be experimentally checked up to which extent and under which conditions the terminal velocity of body *A* is 4 cm/s. Thus,

$$M_1 = M_2 = M,\ v_1 = 4\text{ cm/s} \tag{9.39}$$

Now it can be understood under what experimental conditions (compositions of bodies, point of impact, nature of resistive forces etc.) the terminal velocity of body *A* equals the initial velocity of body *B*, i.e. 4 cm/s.

The terminal velocity of body *B*, i.e. $v_2$ can be calculated from Eq. (9.27) by substituting various values

$$v_2 = \frac{(M_2 - M_1)u_2 + 2M_1u_1}{M_1 + M_2} = \frac{(M - M)u_2 + 2Mu_1}{M + M_2} = \frac{0 + 2Mu_1}{2M}$$

$$v_2 = u_1 = 10 \text{ cm/s}$$

Terminal velocity of body $B$ = Initial velocity of body $A$ = 10 cm/s

Thus, the terminal velocity of body $B$ ($v_2$) is precisely equal to 10 cm/s. It can be experimentally checked up to which extent and under which conditions the terminal velocity of body $B$ is 10 cm/s. Thus,

$$M_1 = M_2 = M,\ v_2 = 10 \text{ cm/s}$$

Now it can be understood under what experimental conditions (compositions of bodies, point of impact, nature of resistive forces), the terminal velocity of body $B$ equals the initial velocity of body $A$. In the existing literature there is no experimental data which may justify the above calculations. As the collision is elastic, so the coefficient of restitution is unity. Thus, the value of $e$ can be calculated from Eq. (9.17)

$$\text{Coefficient of restitution, } e = \frac{v_2 - v_1}{u_1 - u_2} = \frac{(10 - 4) \text{ cm/s}}{(10 - 4) \text{ cm/s}} = 1$$

### *5.1(b). When target is at rest ($u_2 = 0$)*

When the target body is at rest, i.e. $u_2 = 0$.

In Eq. (9.25) and Eq. (9.27) setting $u_2 = 0$, we have

$$v_1 = \frac{(M_1 - M_2)u_1}{(M_1 + M_2)} \tag{9.30}$$

$$v_2 = \frac{2M_1u_1}{M_1 + M_2} \tag{9.31}$$

Various sub-cases of Eqs. (9.30) and (9.31) can be considered.

(a) **When two bodies are of equal masses:**
Setting $M_1 = M_2 = M$, $u_1 = 10$ cm/s. Under this condition in Eq. (9.30) and Eq. (9.31) we get

$$v_1 = \frac{(M - M)u_1}{(M + M)} = 0$$

Here the basic assumption is that the target body $B$ is at rest ($u_2 = 0$) and the bodies are of equal masses ($M_1 = M_2 = M$).

Thus, the body $A$ collides with the body $B$ of equal mass and then comes to rest ($v_1 = 0$).

$$v_2 = \frac{2M_1u_1}{M_1+M_2} = \frac{2Mu}{M+M} = u_1 \tag{9.33}$$

or $$v_2 = u_1 = 10 \text{ cm/s}$$

Thus, under this condition the target body $B$ will move with the initial velocity of body $A$, i.e. 10 cm/s and the projectile body $A$ comes to rest. The experiment needs to be conducted by hitting soft cloth or cotton ball on the steel or wooden ball. Now it can be experimentally established under which conditions the soft cloth ball comes to rest ($v_1 = 0$) and the steel ball moves with the velocity of the soft ball (10 cm/s). Obviously, it is not true for every medium having varied ranges of resistive forces. The experiments can be conducted by replacing the soft cloth ball with the steel ball of the same mass. As the collision is elastic so the coefficient of restitution is unity. Thus the value of $e$ can be calculated from Eq. (9.17)

$$\text{Coefficient of restitution, } e = \frac{v_2 - v_1}{u_1 - u_2} = \frac{(10-0)\text{cm/s}}{(10-0)\text{cm/s}} = 1$$

(b) When the mass of body $B$ is of negligible mass compared to body $A$, it implies that

$$M_2 << M_1 \text{ or } M_1 + M_2 = M_1 \text{ and } M_1 - M_2 = M_1$$

Now substituting the various values ($M_1 + M_2 = M_1$) in Eq. (9.30) we get

$$v_1 = \frac{M_1u_1}{M_1} = u_1 = 10 \text{ cm/s} \tag{9.40}$$

This case is mathematically feasible but it would be hard to realize it experimentally. Thus body $A$ will keep on moving with its original velocity of 10 cm/s. Now the terminal velocity of the target can be calculated by substituting various values in Eq. (9.31)

$$v_2 = \frac{2M_1u_1}{M_1} = 2u_1 = 10 \times 2 = 20 \text{ cm/s} \tag{9.41}$$

Thus, if the target or body $B$ is of negligible mass then body $B$ will move with twice the velocity of body $A$ (projectile). As the collision is elastic, so the coefficient of restitution is unity. Thus the value of $e$ can be calculated from Eq. (9.17)

$$\text{Coefficient of restitution, } e = \frac{v_2 - v_1}{u_1 - u_2} = \frac{(20-10)\ \text{cm/s}}{(10-0)\ \text{cm/s}} = 1$$

Now the conditions under which predictions are justified can be experimentally checked.

It is mentioned that

$$\text{Mass of body } B << \text{mass of body } A \text{ or } M_2 << M_1$$

But is a qualitative perception only. It would be better to point out exactly how smaller the body $B$ is compared with body $A$.

Mass of body $B$ = 0.01 mass of body $A$
Mass of body $B$ = 0.001 mass of body A

The mass of body $B$ = 0.0001 mass of body $A$. It would be prudent to explore the conditions under which the velocity of body $B$ may be $nu_2$ ($n$ is different from 2)

So velocity has to be measured with the utmost care in such macroscopic level experiments.

### 5.2. *Large difference in velocities of body A and body B*

We have considered that the velocity of body $A$ is 10 cm/s and that of body $B$ is 4 cm/s. As the velocity of body $A$ is more than that of body B, hence collisions occur in the following cases.

(i) The velocity of body $A$ is 4.1 cm/s and that of body $B$ 4 cm/s.
(ii) The velocity of body $A$ is 40 cm/s and that of body $B$ 4 cm/s.
(iii) The velocity of body $A$ is 4000 cm/s and that of body $B$ 4 cm/s.

Thus, in the first case the difference in the velocity of body $A$ and body $B$ is 0.1 cm and the same in the third case is 3996 cm/s. Now it can be speculated whether the difference in the velocities of bodies can play any role in elastic collisions. It can be confirmed experimentally. The difference in velocity can be more than the above mentioned value. Bodies of different composition can also be considered.

### 5.3. *Concluding remarks*

Thus, these equations of elastic collisions involve the conservation of momentum and the conservation of kinetic energy at the macroscopic level. In all the above calculations the coefficient of restitution is also unity, which is another condition for the collision to be elastic. Now it is obvious that in the mathematical equations there are no factors which may take into account the characteristics of bodies and the resistive system. However, in this nature of composition of bodies, the resistive forces, the point of impact may play significant roles. As these are basic laws of physics, so it is stressed that the experiments based upon the above equation must be conducted to check the predictions at the macroscopic level. It will help in understanding the conservation laws at the macroscopic level. A possible experimental set up is described as below.

## 6.0. Experimental aspects

Resnick [1] mentioned that the collisions between large bodies are always inelastic to some extent; however, the collisions between ivory or glass balls may be regarded as approximately elastic. Such comments put constraints on the conservation law at the macroscopic level. It implies that in one dimension at the macroscopic level collisions (between gross bodies) will be perfectly elastic under some rigorous conditions only. The aim of the investigations is to find out the most suitable experimental conditions so that the equations based on the one-dimensional collisions are precisely obeyed. Apparently the collisions are precisely elastic under certain conditions only. The following experimental set up may be used in the confirmation of above equations. However, improvements in the experimental set up are wide open for proper conductance of experiments.

**(a)** **Frame:** For quantitative experimental observations a frame with the following characteristics may have to be fabricated. The graduate frame of a suitable material may have a length of 10 m and a width and height of 0.5 m. The frame should have a window to put suitable spheres (described below) in it and be fitted with a vacuum pump to minimize air resistance.

The resistive forces (frictional force, air resistance, adhesive forces, etc.) as far as possible should not affect the uniform motion of spheres (which the body inherits to maintain by the virtue of the property of inertia of motion). Consequently, while choosing the material of the base and spheres, the characteristics of the combination should be such that the coefficients of static and dynamic friction should be the least. Thus, in this regard the base of the frame may be teflon and spheres of steel or teflon as coefficients of static and dynamic friction (for teflon–steel or teflon–teflon bodies) are 0.04 each [2]. However, typical observations may made for systems having higher values, e.g. the coefficients of static friction is 0.74 and that of dynamic friction 0.57 for the steel body on the surface of steel.

In addition to the frame should have a suitable room for spokes and motors arrangement (as described below), preferably at the roof of the frame. The frame should be placed at the perfectly smooth platform without inclination on either side. The dimensions of the frame may vary depending upon moving spheres (bodies) and general suitability of experiment.

(ii) **Spheres:** The equations of elastic collisions are meant for all moving bodies but can be confirmed for the first instance for spheres for experimental simplicity. Thus the spheres of homogeneous and suitable composition of steel or teflon have to be designed as already mentioned so that the coefficients of static and dynamic friction are quite less for these comparatively. Also spheres of other material can be chosen for simplicity. The outer surface of the sphere should be devoid of all types of irregularities of deformations, and the internal composition of spheres should be perfectly uniform as far as possible.

(iii) **Spokes and motors:** The spheres have to be simultaneously pushed with suitable initial velocities such that $u_1 > u_2$. Thus two spokes should be fitted on the roof of the frame attached to electric motors each having an independent and suitably regulated power supply. The working of spokes is similar to the blades of a fan but in a much more controlled way. The spoke should hit the spheres perfectly in the middle so that spheres move initially along the line joining their centres, then after the head-on collision the spheres should move along the same straight line. Resnick and Halliday [1] put constraints on moving spheres that they should be non-rotating before and after the collision (this conditions may be easily obtained if blocks are used instead of the spheres) but have not discussed the case (even theoretically) if the spheres rotate. This may be possibly due to reason that the rotating spheres may not precisely move in a straight line. In addition, the spheres and the blocks (if used) should not be deformed when the spokes hit them during collisions.

(iv) **An interferometer in conjunction with a quartz clock:** It is very important to measure the velocities (displacements in particular time) of the spheres accurately. The displacement of the spheres (or blocks if used) can be measured with the help of an interferometer and time with a quartz clock used in conjunction with the interferometer. This method is well discussed in the existing literature, as widely utilized in the measurement of the acceleration due to gravity [3], upward and downward speeds of small spheres in the determination of the viscosity of fluids [4]. This arrangement can be judiciously set along the walls or at the roof of the frame. Here in the determination of velocities of the spheres the principle can be the same as used in measuring the speed of a tennis ball.

The basic requirement of the quantitative confirmation of elastic collisions is that constant velocities of spheres should be sustained at least during the times of observations. The presence of gravity hinders the perpetual uniform motion of the bodies (spheres), even if system is ideally devoid of resistive forces. While in motion the energy may be dissipated [5] in various forms, it should be minimized and be estimated. Further it is added that it has been the theoretical description of the experimental setup. So depending upon the experimental suitability and technical improvement, some alternations in it may definitely be made so that the described apparatus is more useful. Finally, it is added that the above mathematical equations must be put to experimental tests so that final conclusions about the equations may be drawn [6]. The importance of such experiments lies in the fact that conservation laws are checked at the macroscopic level.

## Frequently asked questions

1. **Can the mathematical equations based upon elastic collisions confirm the conservation laws?**

   **Answer:** Yes, these equations are based upon the law of conservation of energy and momentum. If these equations are experimentally confirmed then the conservation law will also be justified by this macroscopic method. Apparently the laws will be confirmed under some conditions due to the presence of resistive forces in the system and compositions of bodies. So these mathematical equations definitely provide a conceptual basis for the confirmation of conservation laws. The perfect beauty lies in the experimental confirmation of equations. These equations are equally applicable if the experiments are performed in outer space or on the surface of the Moon. For completeness these equations can be analyzed on the heavenly body having acceleration due to gravity more than *g*.

2. **Is the motion of projectiles and targets after collision accelerated or non-accelerated?**

**Answer:** The final velocities of the projectile and the target are easily calculated. The terminal velocity of the projectile is

or $$v_1 = \frac{(M_1 - M_2)u_1 + 2M_2u_2}{(M_1 + M_2)} \tag{9.25}$$

Similarly, the terminal velocity of the target is

$$v_2 = \frac{(M_2 - M_1)u_2 + 2M_1u_1}{M_1 + M_2}$$

Now in collision the masses and initial velocities of projectiles are constant and hence the final velocities are also constant. Thus acceleration is zero, i.e. $a = 0$. Then the kinematical equations ($v = u + at$, $v^2 - u^2 = 2aS$, $S = ut + \frac{1}{2}at^2$, involving acceleration are not applicable. The distance travelled can only be determined by $S = vt$.

3. **Can the recoil of the gun justify the conservation of momentum?**

**Answer:** Yes, it can be justified on the basis of the law of conservation of momentum. Consider a shell of mass 0.01 kg is fired from a gun of mass 25 kg. If the muzzle velocity of the shell is 200 m s$^{-1}$, what is the recoil velocity of the gun?

Applying the law of conservation of momentum, the recoil speed turns out to be

$$0 = 0.01 \text{ kg} \times 200 + 25 \times V$$

$$V = -\frac{0.01 \times 200}{25} = -0.08 \text{ m/s or } -8\text{cm/s}$$

Sensitive experiments are required to check the prediction, i.e. does the gun recoils with a velocity of 8 cm/s or not? In this case the resistive forces play a significant role.

4. **Will it mean the break down of conservation laws if experiments do not succeed at the macroscopic level?**
   **Answer:** When the equations are theoretically derived then normally ideal conditions are considered. For example in this case the compositions of spheres, resistive forces, point of impact, etc. are not properly taken into account or defined in the mathematical equations. There is a huge difference between the mathematical perceptions and the realistic experimental setup. So under all conditions the equations may not be confirmed but when external factors are controlled the conservation laws at the macroscopic level will be valid. It is difficult to understand why scientists have not seriously performed such experiments which are based upon basic conservation laws. No mathematical equation is above the experiments.

## References


1. Resnick, R. and Halliday, D., Phys. Part I (Willey Eastern Limited, New Delhi, Forty-Second Reprint), pp. 5, 215–218 (1992).
2. Sears, W.F., Zemansky, M.W., and Young, H.D., University Phys. (Narosa Publishing House, New Delhi, 6th Edition), pp. 30–31, pp. 148–154 (1994).
3. McGraw Hill Encyclopeadia of Science and Technology, (McGraw-Hill Book Company, New York.). No. 4, pp. 161–162. No. 5, pp. 482 and No. 9, pp. 289 (1987).
4. Sharma, A., Acta Ciencia Indica, Phys. Vol. XXII, pp. 22–24 (1996).
5. Green, C.R., Tech. Phys. (Prentice-Hall, Inc. Englewood Cliffs, New Jersey), pp. 124–131 (1984).
6. Sharma, A., Acta Ciencia Indica, Phys. Vol. XXIV P, No. 4, pp. 127–130 (1998).

10

# One-Dimensional Elastic Collisions and Newton's Third Law of Motion

**First glimpse**

- ✓ Newton quoted three laws of motion in the Principia very briefly.
- ✓ These laws are quoted at page nos. 19–20 of the first English translation of The Principia by Andrew Mott in 1729.
- ✓ The third law of motion is understood qualitatively at the macroscopic level in standard textbooks.
- ✓ The third law of motion can be justified on the basis of equations based upon elastic collisions, especially when the target is very-2 heavy than the projectile.
- ✓ The real beauty of science is to experimentally confirm the mathematical equations.
- ✓ In view of discussion it is concluded that
  'To every action there is opposite reaction, but the reaction may or may be always equal to action depending upon the characteristics of the system'.

## 1.0. The third law of motion as in The Principia

Sir Isaac Newton published the Philosophiae Naturalis Principia Mathematica (Mathematical Principles of Natural Philosophy), in Latin in 1687 and its second and third editions appeared in 1713 and 1726. It was translated by Andrew Mott in 1729 into English, which is referred here in regard to the third law of motion. In this masterpiece at pages 19 and 20 Newton quoted three laws of motion very briefly. The original description of Newton's third law of motion is given here. For example, Newton's third law of motion is explained in just two paragraphs as quoted below. Our discussion is based upon this original law and explanation as given in the Principia. More discussion is included here which is not taken into account by Newton, e.g. discussion on the relevant results of elastic collision. In elastic collision the body recoils thus the reaction occurs opposite to the action. Here Newton's third law of motion [1] is quoted both in original Latin and its English translation.

The original form of Newton's third law of motion as in Latin:

***Lex III:*** *Actioni contrariam semper et æqualem esse reactionem: sive corporum duorum actiones in se mutuo semper esse æquales et in partes contrarias dirigi.*

When translated into English

**Law III:** To every action there is always an equal and opposite reaction: or the forces of two bodies on each other are always equal and are directed in opposite directions.

In the Principia Sir Isaac Newton explained the third law of motion in the following way. However, different scientists and authors tried to explain the same in their own way but the soul of the law is intact. Some have given quantitative proofs and some have explained the law qualitatively only. It is clear from the existing literature that the law is understood qualitatively at the macroscopic level.

### *1.1. Newton's original explanation to third law of motion*

Whatever draws or presses another is as much drawn or pressed by the other. If you press a stone with your finger, the finger is also pressed by the stone. If a horse draws a stone tied to a rope, the horse (if I may so say) will be equally drawn back towards the stone: for the distended rope, by the same endeavour to relax or unbend itself, will draw the horse as much towards the stone as it does the stone towards the horse, and will obstruct the progress of the one as much as it advances that of the other.

If a body impinges upon another, and by its force change the motion of the other, that body also (became of the quality of the mutual pressure) will undergo an equal change, in its own motion, towards the contrary part. The changes made by these actions are equal, not in the velocities but in the motions of bodies; that is to say, if the bodies are not hindered by any other impediments. For, because the motions are equally changed, the changes of the velocities made towards contrary parts are reciprocally proportional to the bodies. This law takes place also in attractions, as will be proved in the next scholium.

### *1.2. Examples which illustrate the law qualitatively*

**(a)** **Swimming in water:** Consider a swimmer in diving in the pool of water. *A* swimmer pushes the water in backward direction. It may be regarded as action. The water pushes the swimmer in forward direction with the same force, it is reaction. As both action and reaction are equal (according to the third law of motion), hence the swimmer moves forward in water. In this case both the body which applies action and the medium which imparts reaction are in motion.

However, for proper understanding the magnitudes of action and reaction have to be measured. Here the action may be taken as the force which swimmer applies on the water, and the reaction as the force which water exerts on the man. Here,

Action = Force applied by the swimmer on water
Reaction = Force applied by water on the swimmer

To draw the quantitative conclusions, both are to be precisely measured.

**(b)** **Flight of jet planes and rockets:** When a rocket moves forward in space the fuel is burnt and escapes backward through the nozzle in form of highly compressed gases. It is action in terms of Newton's third law of motion. The escaping gases push the jet plane or rocket forward with the same force (reaction). Hence jet or rocket moves forward. From here it is concluded that in this case backward speed of escaping gases must be quiet high as heavy jet or rocket moves forward with high velocity.

It must be regarded as a qualitative explanation. Here the action would be the backward momentum of the gases coming from nozzle in the form of highly compressed gases whereas the reaction would be the forward momentum of the rocket.

Action = Momentum of the highly compressed gases

Reaction = Momentum (forward) of the rocket

To draw the quantitative conclusions, both are to be precisely measured, the action must be equal to that of the reaction.

(c) **It is difficult to drive a nail into a wooden block without holding the block**
Firstly, consider the case when a wooden block is not placed against a solid support. When a nail is driven into it, the block moves and, as a result, the nail cannot pierce the wooden block. In this cases the reaction (provided by block) is smaller. The external force applied on the nail is termed as action. When the block is placed against the solid support (say on the floor), and hit with hammer the block gets the reaction. Thus, the nail is driven into the wooden block. However, the nail can be driven into a watermelon even when held on the hand. So the reaction also depends upon the nature and characteristic of the body. To understand the phenomena quantitatively, the exact values of action and reaction are to be measured.

(d) **It is difficult to walk on the sand:** when we move forward, the ground is pushed with the feet and this is called action. If a similar reaction is obtained from the ground we are able to walk. However when we try to walk on the sand, the sand itself is pushed down without generating the adequate amount of reaction. Thus it is difficult to walk on the sand. However, in this case it is difficult to precisely measure the action exerted by the person on ground with the feet and the reaction of the sand. Until quantitative measurements are taken the explanation must be regarded as qualitative.

(e) **Rebounding on the rubber wall:** When a rubber ball is struck against the wall, it exerts a force on the wall as action. The ball will rebound with the equal force (reaction) exerted by the wall on the ball. It is believed that the ball rebounds due to the reaction. In this case, both action and reaction are equal, according to Newton's second law of motion. The critical analysis of this phenomenon is discussed below.

### *1.3. The following example is neglected*

Consider a boy is standing in front of a concrete wall, holding a rubber ball and a cloth ball in the hands. Let the wall is at the distance of 10 m from the boy.

**First case:** Let the boy throws the rubber ball at the wall with force $F$. The ball strikes the wall, and comes back to the boy, i.e. travelling a distance of 10 m. Now action and reaction can be understood as

Action: the boy pushes the ball towards the wall
Reaction: the wall pushes the ball towards the boy.

As the boy pushed the ball from distance of 10 m and the wall rebounds to a distance of 10 m so the action and the reaction are equal and opposite.

The boy pushes the ball through 10 m = the wall rebounds the ball through 10 m

$$\text{Action} = -\text{Reaction} \qquad (10.1)$$

The $-ve$ sign indicates that after the reaction the ball moves in the opposite direction. In this case, Newton's third law of motion is completely obeyed.

**Second case:** Let the boy throws the cloth ball at the wall which is at a distance of 10 m.

Let the boy throws the cloth wall with the same force $F$. The ball strikes the wall and rebounds to only 5 m. Now action and reaction can be understood as

Action: the boy pushes the ball towards the wall
Reaction: the wall pushes the ball towards the boy

So the boy pushed the cloth ball from a distance of 10 m and the wall returns to a distance of 5 m.

The boy pushes the ball towards the wall through 10 m ≠ The wall rebounds the ball through 5 m

$$\text{Action} \neq \text{Reaction}$$

So the action and the reaction are opposite but not equal in the case of the cloth ball. The reaction also depends upon various characteristics of the body or the ball (cloth, rubber) and wall (concrete or wooden or earthen). Thus, the reaction depends upon various characteristics of the system. It is just possible that the characteristics of the ball and the wall are such that the ball may rebound more than 10 m. For further understanding a cushioned wall can be considered for collision which will affect reaction considerably. For the same action as in previous case the reaction would be less.

## 3.0. Elastic collision in one dimension: mathematical equations

The collision in which both the momentum and kinetic energy are conserved and the colliding bodies continue to move in the same straight line after collision, is elastic collision in one dimension.

Consider two perfectly elastic bodies $A$ (projectile) and $B$ (target) of masses $M_1$ and $M_2$ moving along the same straight line with velocities $u_1$ and $u_2$ respectively. The bodies will collide only if $u_1 > u_2$.

The two bodies undergo a head-on collision and continue moving along the same straight line with velocities $v_1$ and $v_2$ in the same direction. Two bodies will separate after collisions only if $v_2 > v_1$.

As in an inelastic collision, the momentum is conserved and we have

$$M_1u_1 + M_2u_2 = M_1v_1 + M_2v_2 \tag{10.2}$$

Since kinetic energy is also conserved, we have

$$\frac{M_1u_1^2}{2} + \frac{M_2u_2^2}{2} = \frac{M_1v_1^2}{2} + \frac{M_2v_2^2}{2} \tag{10.3}$$

From Eq. (10.2) we have

$$M_1(u_1 - v_1) = M_2(v_2 - u_2) \tag{10.4}$$

Also from Eq. (10.3) we can write

$$M_1(u_1^2 - v_1^2) = M_2(v_2^2 - u_2^2) \tag{10.5}$$

Dividing Eq. (10.5) by Eq. (10.4)

$$\frac{\left(u^2{}_1 - v_1^2\right)}{\left(u_1 - v_1\right)} = \frac{\left(v_2^2 - u_2^2\right)}{\left(v_2 - u_2\right)}$$

$$u_1 + v_1 = v_2 + u_2$$

$$u_1 - u_2 = v_1 + u_2 \tag{10.6}$$

Dividing both sides of Eq. (10.6) by $u_1 - u_2$

$$e = \frac{v_2 - v_1}{u_1 - u_2} \tag{10.7}$$

where $e$ is the coefficient of restitution or the coefficient of resilience and for an elastic collision is unity.

Let us find the velocity of body $A$ after collision. From Eq. (10.6) we have

$$v_2 = u_1 - u_2 + u_2$$

Substituting $v_2$ in Eq. (10.4), we get

$$\begin{aligned} M_1u_1 + M_2u_2 &= M_1v_1 + M_2(u_1 - u_2 + v_1) \\ &= M_1v_1 + M_2u_1 - M_2u_2 + M_2v_1 \\ (M_1 - M_2)u_1 + 2M_2u_2 &= (M_1 + M_2)v_1 \end{aligned}$$

Or $$v_1 = \frac{(M_1 - M_2)u_1 + 2M_2u_2}{(M_1 + M_2)} \tag{10.8}$$

Again from Eq. (10.6)

$$v_1 = v_2 - u_1 + u_2 \tag{10.9}$$

Substituting $v_1$ from Eq. (10.9) into Eq. (10.2)

$$\begin{aligned} M_1u_1 + M_2u_2 &= M_1(v_2 - u_1 + u_2) + M_2v_2 \\ &= M_1v_2 - M_1v_1 - M_1u_2 + M_2v_2 \\ (M_2 - M_1)u_2 + 2M_1u_1 &= (M_1 + M_2)v_2 \end{aligned}$$

or $$v_2 = \frac{(M_2 - M_1)u_2 + 2M_1u_1}{M_1 + M_2} \tag{10.10}$$

The known value of initial velocities can be imparted to body $A$ and body $B$ with sensitive equipment. The values of $v_1$ and $v_2$ can be calculated with the help of Eq. (10.8) and Eq. (10.19). Let us calculate the final velocities of two bodies ($v_1$, $v_2$) after a collision in the following special cases.

## 4.0. When body $B$ is at rest ($u_2 = 0$)

Now substituting the value of $u_2 = 0$ in Eq. (10.8) and Eq. (10.10)

$$v_1 = \frac{(M_1 - M_2)u_1}{(M_1 + M_2)} \tag{10.11}$$

$$v_2 = \frac{2M_1u_1}{M_1 + M_2} \tag{10.12}$$

These simplified equations, based upon one-dimensional elastic collisions, will now be used for understanding of Newton's third law of motion.

### *4.1. Mass of body B is very-2 large compared to mass of body A*

Let us consider that the mass of body *B* (target) is much larger than that of body *A* (projectile). The body *A* strikes body *B* and rebounds. The various cases are discussed below.

(i) When $M_2 >> M_1$ i.e. the target (body *B*) is very-2 heavy compared to the projectile (body *A*).

Thus
$$M_1 - M_2 = -M_2,$$
$$M_2 + M_1 = M_2$$

Let the projectile moves with initial velocity $u_1$ which may be regarded as action. Thus

$$\text{Action} = u_1$$

Let after collision with the target the projectile moves with velocity $v_1$. Thus

$$\text{Reaction} = v_1$$

In this case the final velocities of the projectile and the target can be calculated from Eq. (10.11) and Eq. (10.12).

$$v_1 = \frac{-M_2 u_1}{M_2} = -u_1 \tag{10.13}$$

The negative sign in $v_1$ indicates that the direction of body *A* after collision reverses.

$$v_2 = \frac{2M_1}{M_2} u_2$$

As $M_2$ is very-2 large compared to $M_1$, thus ratio $M_1/M_2$ will tend to zero.

$$v_2 \approx 0 \tag{10.14}$$

Further body *B* moves with an infinitesimally small velocity. In view of the third law of motion

$$\text{Action} = u_1 \tag{10.15}$$
$$\text{Reaction} = -u_1 \tag{10.16}$$

The negative sign indicates that the direction of the projectile becomes opposite after the collision, i.e. it starts moving backward. Thus, if the forward movement of the body *A* is the action then reverse movement is the reaction.

In the cases discussed earlier in Chapter 9, the direction of body $A$ before and after the collision is the same. When a ball hits the wall it rebounds obeying this case. The magnitudes of action and reaction are precisely equal in Eq. (10.13).

Thus in this case ($M_2 >> M_1$ or body $B >>$ body $A$ or target >> projectile) the reaction is precisely EQUAL to the action and OPPOSITE in direction. Hence, in this case

$$\text{Magnitude of Action } (u_1) = \text{Magnitude of Reaction } (u_1) \tag{10.17}$$

Direction : Opposite (indicated by the negative sign)

$$\text{Coefficient of restitution, } e = \frac{v_2 - v_1}{u_1 - u_2}$$

$$= [0-(-u_1)]/[u_1 - 0] = 1$$

(ii) Consider the case when the target (body $B$) is assumed to 1000 times larger than the projectile (body $A$). Thus,

$$M_1 - M_2 = M_1 - 1000M_1 = -999M_1,$$
$$M_1 + M_2 = M_1 + 1000M_1 = 1001M_1$$

Let the projectile moves with initial velocity $u_l$ which may be regarded as action. Thus

$$\text{Action} = u_1$$

Let after the collision with the target the projectile moves with velocity $v_1$. Thus

$$\text{Reaction} = v_1$$

In this case the final velocities of the projectile and the target can be calculated from Eq. (10.11) and Eq. (10.12).

$$v_1 = \frac{2M_1u_1}{(M_1 + 1000M_1)} = \frac{-999M_1u_1}{1001M_1} = -0.998001998u_1 \tag{10.18}$$

The negative sign in $v_1$ indicates that the direction of body $A$ after the collision reverses.

$$v_2 = \frac{2M_1u_1}{(M_1 + 1000M_1)} = \frac{2M_1u_1}{1001M_1} = 0.001998001u_1 \tag{10.19}$$

In view of the third law of motion

$$\text{Action} = u_1 \tag{10.15}$$

$$\text{Reaction} = -0.998001998u_1 \tag{10.20}$$

The negative sign indicates that the direction of the projectile becomes opposite after the collision, i.e. it starts moving backward. Thus, if the forward movement of the body *A* is the action then the reverse movement is the reaction. In the cases discussed earlier in Chapter 9, the direction of body *A* before and after collision is the same. When the ball hits the wall, it rebounds obeying this case. The magnitudes of action and reaction are precisely equal.

Thus, in this case ($M_2 = 1000M_1$, $u_2 = 0$) action and reaction are not equal but opposite in direction.

$$\text{Magnitude of action } (u_1) \neq \text{Magnitude of reaction } (0.998001998u_1)$$

Direction: opposite (indicated by the negative sign).

Thus, Newton's third law of motion is not completely obeyed here, as the magnitudes of action and reaction are different.

$$\text{Coefficient of restitution, } e = \frac{v_2 - v_1}{u_1 - u_2}$$

$$\left[\frac{2M_1u_1}{1001M_1} - \frac{-999M_1u_1}{1001M_1}\right]/[u_1 - 0] = 1$$

(iii) Consider the case when body *B* (target) is assumed to be 500 times larger than body *A* (projectile). Thus,

$$M_1 - M_2 = M_1 - 500M_1 = -499M_1,$$
$$M_1 + M_2 = M_1 + 500M_1 = 501M_1$$

Let the projectile move with initial velocity $u_1$ which may be regarded as action. Thus

$$\text{Action} = u_1$$

Let after collision with target the projectile moves with velocity $v_1$. Thus

$$\text{Reaction} = v_1$$

In this case, the final velocities of the projectile and the target can be calculated from Eq. (10.11) and Eq. (10.12).

$$v_1 = \frac{(M_1 - 500M_1)u_1}{(M_1 + 500M_1)} = \frac{-499M_1u_1}{501M_1} = -0.9960798u_1 \tag{10.21}$$

The negative sign in $v_1$ indicates that the direction of body $A$ after collision reverses.

$$v_2 = \frac{2M_1u_1}{(M_1 + 500M_1)} = \frac{2M_1u_1}{501M_1} = 0.003992015u_1 \qquad (10.22)$$

In view of the third law of motion

$$\text{Action} = u_1 \qquad (10.15)$$

$$\text{Reaction} = -0.9960798u_1 \qquad (10.23)$$

The negative sign indicates that the direction of the projectile becomes opposite after the collision, i.e. it starts moving backward. The magnitudes of action and reaction are not precisely equal.

Thus, in this case ($M_2 = 500M_1$, $u_2 = 0$) the action and the reaction are not equal but opposite in direction.

Magnitude of action ($u_1$) $\neq$ Magnitude of reaction ($0.980198u_1$)

Direction: opposite (indicated by negative sign).

Thus, Newton's third law of motion is not completely obeyed here, as the magnitudes of action and reaction are different.

(iv) Consider the case when the target (body $B$) is assumed to be 100 times larger than the projectile (body $A$). Thus,

$$M_1 - M_2 = M_1 - 100M_1 = -99M_1,$$
$$M_1 + M_2 = M_1 + 100M_1 = 101M_1$$

Let the projectile moves with initial velocity $u_1$ which may be regarded as action. Thus

$$\text{Action} = u_1$$

Let after the collision with the target the projectile moves with velocity $v_1$. Thus

$$\text{Reaction} = v_1$$

In this case the final velocities of the projectile and the target are calculated from Eq. (10.11) and Eq. (10.12).

$$v_1 = \frac{(M_1 - 100M_1)u_1}{(M_1 + 100M_1)} = \frac{-99M_1u_1}{101M_1} = -0.980198u_1 \qquad (10.24)$$

The negative sign in $v_1$ indicates that the direction of body $A$ after the collision reverses.

$$v_2 = \frac{2M_1u_1}{(M_1 + 100M_1)} = \frac{2M_1u_1}{101M_1} = 0.01980198u_1 \qquad (10.25)$$

In view of the third law of motion

$$\text{Action} = u_1 \qquad (10.15)$$
$$\text{Reaction} = -0.980198u_1 \qquad (10.26)$$

The negative sign indicates that the direction of the projectile becomes opposite after the collision, i.e. it starts moving backward. Thus, if the forward movement of the body *A* is action then the reverse movement is reaction. In the cases discussed earlier in Chapter 9, the direction of body *A* before and after collision is the same. When the ball hits the wall it rebounds obeying this case. The magnitudes of action and reaction are not precisely equal.

Thus, in this case ($M_2 = 100M_1$, $u_2 = 0$) action and reaction are not equal but opposite in direction.

Magnitude of action ($u_1$) ≠ Magnitude of reaction ($0.980198u1$)

Direction: Opposite (indicated by negative sign)

Thus, Newton's third law of motion is not completely obeyed here as the magnitudes of action and reaction are different.

The coefficient of restitution in this case is also unity.

(v) Consider the case when body *B* (target) is assumed to be 10 times larger than body *A* (projectile). Thus,

$$M_1 - M_2 = M_1 - 10M_1 = -9M_1,$$
$$M_1 + M_2 = M_1 + 10M_1 = 11M_1$$

Let the projectile moves with initial velocity $u_1$ which may be regarded as action. Thus

$$\text{Action} = u_1$$

Let after collision with the target the projectile moves with velocity $v_1$. Thus

$$\text{Reaction} = v_1$$

In this case the final velocities of the projectile and the target can be calculated from Eq. (10.12) and Eq. (10.12).

$$v_1 = \frac{(M_1 - 10M_1)u_1}{(M_1 + 10M_1)} = \frac{-9M_1u_1}{11M_1} = -0.818181u_1 \tag{10.27}$$

The negative sign in $v_1$ indicates that the direction of body $A$ after the collision reverses.

$$v_2 = \frac{2M_1u_1}{(M_1 + 10M_1)} = \frac{2M_1u_1}{11M_1} = 0.181818u_1 \tag{10.28}$$

In view of the third law of motion

$$\text{Action} = u_1 \tag{10.15}$$

$$\text{Reaction} = -0.818181u_1 \tag{10.29}$$

The negative sign indicates that the direction of the projectile becomes opposite after the collision, i.e. it starts moving backward. The magnitudes of action and reaction are not precisely equal.

Thus, in this case ($M_2 = 10M_1$, $u_2 = 0$) action and reaction are not equal but opposite in direction.

Magnitude of action ($u_1$) ≠ Magnitude of reaction ($0.818181u_1$)

Direction: opposite (indicated by negative sign)

Thus, Newton's third law of motion is not completely obeyed here, as the magnitudes of action and reaction are different.

(vi) Consider the case when the target (body $B$) is assumed to be 5 times larger than the projectile (body $A$), $M_2 = 5M_1$.

Thus,

$$M_1 - M_2 = M_1 - 5M_1 = -4M_1,$$
$$M_1 + M_2 = M_1 + 5M_1 = 6M_1$$

Let the projectile move with initial velocity $u_1$ which may be regarded as action. Thus

$$\text{Action} = u_1$$

Let after collision with the target the projectile (Body $A$) move with velocity $v_1$. Thus

$$\text{Reaction} = v_1$$

In this case the final velocities of the projectile and the target can be calculated from Eq. (10.11) and Eq. (10.12).

$$v_1 = \frac{(M_1 - 5M_1)u_1}{(M_1 + 5M_1)} = \frac{-4M_1u_1}{6M_1} = -0.66666u_1 \qquad (10.30)$$

The negative sign in $v_1$ indicates that the direction of body $A$ after the collision reverses.

$$v_2 = \frac{2M_1u_1}{(M_1 + 5M_1)} = \frac{2M_1u_1}{6M_1} = 0.3333333u_1 \qquad (10.31)$$

In view of the third law of motion

$$\text{Action} = u_1$$
$$\text{Reaction} = 0.66666u_1 \qquad (10.32)$$

The negative sign indicates that the direction of the projectile becomes opposite after the collision, i.e. it starts moving backward. The magnitudes of action and reaction are not equal.

Thus, in this case ($M_2 = 5M_l$, $u_2 = 0$) action and reaction are not equal but opposite in direction.

Magnitude of action ($u_1$) ≠ Magnitude of reaction ($0.66666u_1$)

Direction: opposite (indicated by negative sign).

Thus Newton's third law of motion is not completely obeyed here, as the magnitudes of action and reaction are different.

The coefficient of restitution in this case is also unity.

(vii) Consider the case when the target (Body $B$) is assumed to be twice larger than projectile (Body $A$). This is a case often found in practice. Thus,

$$M_1 - M_2 = M_1 - 2M_1 = -M_1,$$
$$M_1 + M_2 = M_1 + 2M_1 = 3M_1$$

Let the projectile moves with initial velocity $u_1$ which may be regarded as action. Thus

$$\text{Action} = u_1$$

Let after collision with the target the projectile moves with velocity $v_1$. Thus

$$\text{Reaction} = v_1$$

In this case the final velocities of the projectile and the target can be calculated from Eq. (10.11) and Eq. (10.12).

$$v_1 = \frac{(M_1 - 2M_1)u_1}{(M_1 + 2M_1)} = \frac{-M_1 u_1}{3M_1} = -0.33333u_1 \tag{10.33}$$

The negative sign in $v_1$ indicates that the direction of body $A$ after collision reverses.

$$v_2 = \frac{2M_1 u_1}{(M_1 + 2M_1)} = \frac{2M_1 u_1}{3M_1} = 0.66666u_1 \tag{10.34}$$

In view of the third law of motion

$$\text{Action} = u_1 \tag{10.15}$$
$$\text{Reaction} = -0.33333u_1 \tag{10.35}$$

The negative sign indicates that the direction of the projectile becomes opposite after the collision, i.e. it starts moving backward. The magnitudes of action and reaction are not precisely equal.

Thus, in this case ($M_2 = 2M_1$, $u_2 = 0$) action and reaction are not equal but opposite in direction.

$$\text{Magnitude of action } (u_1) \neq \text{Magnitude of reaction } (-0.33333u_l)$$

Direction: opposite (indicated by the negative sign)

Thus, Newton's third law of motion is not completely obeyed here as the magnitudes of action and reaction are different.

The coefficient of restitution in this case is also unity.

(viii) Consider the case when the target (Body $B$) is assumed to 1.5 times larger than the projectile (Body $A$). This is a case often found in practice. Thus,

$$M_1 - M_2 = M_1 - 1.5M_1 = -0.5M_1,$$
$$M_1 + M_2 = M_1 + 1.5M_1 = 1.5M_1$$

Let the projectile moves with initial velocity $u_1$ which may be regarded as action. Thus

$$\text{Action} = u_1$$

Let after collision with the target the projectile moves with velocity $v_1$. Thus

$$\text{Reaction} = v_1$$

In this case the final velocities of the projectile and the target can be calculated from Eq. (10.11) and Eq. (10.12).

$$v_1 = \frac{(M_1 - 1.5M_1)u_1}{(M_1 + 1.5M_1)} = \frac{-0.5M_1u_1}{2.5M_1} = -0.2u_1 \quad (10.36)$$

The negative sign in $v_1$ indicates that the direction of body $A$ after collision reverses.

$$v_2 = \frac{2M_1u_1}{(M_1 + 1.5M_1)} = \frac{2M_1u_1}{2.5M_1} = 0.8u_1 \quad (10.37)$$

In view of the third law of motion

$$\text{Action} = u_1 \quad (10.15)$$
$$\text{Reaction} = 0.2u_1 \quad (10.38)$$

The negative sign indicates that the direction of the projectile becomes opposite after the collision, i.e. it starts moving backward. The magnitudes of action and reaction are not precisely equal.

Thus, in this case ($M_2 = 1.5M_1$, $u_2 = 0$) action and reaction are not equal but opposite in direction.

Magnitude of action ($u_1$) ≠ Magnitude of reaction ($0.998001998u_1$)

Direction: opposite (indicated by the negative sign)

Thus, Newton's third law of motion is not completely obeyed here, as the magnitudes of action and reaction are different.

The coefficient of restitution in this case is also unity.

(ix) Consider the case when the target (body $B$) is assumed to be nearly as large as the projectile (body $A$), i.e. $M_2 = 1.01M_1$. This cause is often found in practice. Thus,

$$M_1 - M_2 = M_1 - 1.01M_1 = -0.01M_1,$$
$$M_1 + M_2 = M_1 + 1.01M_1 = 1.01M_1$$

Let the projectile moves with initial velocity $u_1$ which may be regarded as action. Thus

$$\text{Action} = u_1$$

Let after collision with the target the projectile moves with velocity $v_1$. Thus

$$\text{Reaction} = v_1$$

In this case the final velocities of the projectile and the target can be calculated from Eq. (10.11) and Eq. (10.12).

$$v_1 = \frac{(M_1 - 1.01M_1)u_1}{(M_1 + 1.01M_1)} = \frac{-0.01M_1u_1}{2.01M_1} = -0.004975u_1 \quad (10.39)$$

The negative sign in $v_1$ indicates that the direction of body $A$ after collision reverses.

$$v_2 = \frac{2M_1u_1}{(M_1 + 1.01M_1)} = \frac{2M_1u_1}{2.01M_1} = 0.995024u_1 \quad (10.40)$$

In view of the third law of motion

$$\text{Action} = u_1 \quad (10.15)$$
$$\text{Reaction} = -0.004975u_1 \quad (10.41)$$

The negative sign indicates that the direction of projectile becomes opposite after the collision, i.e. it starts moving backward. Thus if the forward movement of the body $A$ is action then the reverse movement is reaction. In the cases discussed earlier in Chapter 9, the direction of body $A$ before and after collision is the same. When a ball hits the wall it rebounds obeying this case. The magnitudes of action and reaction are not precisely equal.

Thus, in this case ($M_2 = 1.01M_1$, $u_2 = 0$) action and reaction are not equal but opposite in direction.

**Table I.** The coefficients of restitution for various cases when the target body($M_2$) is heavy and at rest ($u_2 = 0$) and the projectile ($M_1$) rebounds

| Sr. No. | $M_1$ | $M_2(u_2 = 0)$ | Initial velocity of $M_1$ is $u_1$ | Terminal velocity of $M_1$ is $v_1$ | Terminal velocity of $M_2$ is $v_2$ | Coefficient of restitution $e = (v_2 - v_1)/(u_1 - u_2)$ |
|---|---|---|---|---|---|---|
| 1 | $M_1$ | $M_2 >> M_1$ | $u_1$ | $-u_1$ | $v_2 \approx 0$ | $e = 1$ |
| 2 | $M_1$ | $1000\ M_1$ | $u_1$ | $-0.99800199u_1$ | $v_2 = 0.001998001u_1$ | $e = 1$ |
| 3 | $M_1$ | $500\ M_1$ | $u_1$ | $-0.9960798u_1$ | $v_2 = 0.003992015u_1$ | $e = 1$ |
| 4 | $M_1$ | $100\ M_1$ | $u_1$ | $-0.980198u_1$ | $v_2 = 0.01980198u_1$ | $e = 1$ |
| 5 | $M_1$ | $10\ M_1$ | $u_1$ | $-0.818181u_1$ | $v_2 = 0.181818u_1$ | $e = 1$ |

| Sr. No. | $M_1$ | $M_2(u_2 = 0)$ | Initial velocity of $M_1$ is $u_1$ | Terminal velocity of $M_1$ is $v_1$ | Terminal velocity of $M_2$ is $v_2$ | Coefficient of restitution $e = (v_2 - v_1)/(u_1 - u_2)$ |
|---|---|---|---|---|---|---|
| 6 | $M_1$ | 5 $M_1$ | $u_1$ | $-0.66666u_1$ | $v_2 = 0.3333333u_1$ | $e = 1$ |
| 7 | $M_1$ | 2 $M_1$ | $u_1$ | $-0.33333u_1$ | $v_2 = 0.66666u_1$ | $e = 1$ |
| 8 | $M_1$ | 1.5 $M_1$ | $u_1$ | $-0.2u_1$ | $v_2 = 0.8u_1$ | $e = 1$ |
| 9 | $M_1$ | 1.01 $M_1$ | $u_1$ | $-0.004975u_1$ | $v_2 = 0.995024u_1$ | $e = 1$ |

Note: If fractional values of $u_i$'s and $v_i$'s are taken then e is exactly unity

**Table II.** The initial velocity (action) and terminal velocity (reaction) of the projectile in elastic collisions, and the third law of motion. $u_1$ can be determined experimentally for various bodies

| Sr. No. | $M_1$ | $M_2(u_2 = 0)$ | Action velocity of $M_1$ is $u_1$ | Reaction velocity of $M_1 v_1$ | Comparison | Third law |
|---|---|---|---|---|---|---|
| 1 | $M_1$ | $M_2 >> M_1$ | $u_1$ | $-u_1$ | Equal and opposite | Obeyed |
| 2 | $M_1$ | 1000 $M_1$ | $u_1$ | $0.998001998u_1$ | Unequal and opposite | Not precisely obeyed |
| 3 | $M_1$ | 500 $M_1$ | $u_1$ | $-0.9960798u_1$ | Unequal and opposite | Not precisely obeyed |
| 4 | $M_1$ | 100 $M_1$ | $u_1$ | $-0.980198u_1$ | Unequal and opposite | Not precisely obeyed |
| 5 | $M_1$ | 10 $M_1$ | $u_1$ | $-0.818181u_1$ | Unequal and opposite | Not precisely obeyed |
| 6 | $M_1$ | 5 $M_1$ | $u_1$ | $-0.66666u_1$ | Unequal and opposite | Not precisely obeyed |
| 7 | $M_1$ | 2 $M_1$ | $u_1$ | $-0.33333u_1$ | Unequal and opposite | Not precisely obeyed |
| 8 | $M_1$ | 1.5 $M_1$ | $u_1$ | $-0.2u_1$ | Unequal and opposite | Not precisely obeyed |
| 9 | $M_1$ | 1.01 $M_1$ | $u_1$ | $-0.004975u_1$ | Unequal and opposite | Not precisely obeyed |

Magnitude of action ($u_1$) $\neq$ Magnitude of reaction ($0.004975u_1$)

Direction: opposite (indicated by negative sign)

Thus, Newton's third law of motion is not completely obeyed here, as the magnitudes of action and reaction are different.

The coefficient of restitution in this case is also unity, i.e.

$$e = \frac{v_2 - v_1}{u_1 - u_2} = [\frac{2M_1u_1}{2.01M_1} + \frac{-0.01M_1u_1}{2.01M_1}]/[u_1 - 0] = 1$$

## 5.0. Generalized form of the third law of motion

The above discussion allows us to reassess Newton's third law [2] of motion as:

'To every action there is opposite reaction, but the reaction may or may be equal to the action depending upon the characteristics of the system'.

Thus, the coefficient of restitution and the status of Newton's second law of motion in elastic collision is shown in Table I and Table II

## Frequently asked questions

1. **Newton's third law of motion is explained on the basis of followings:**
   **(a) A swimming of a person**
   **(b) Motion of rocket**
   **(c) Difficult to walk on sand and ice**
   **(d) Rebounding of ball after hitting the wall**

   **Answer:** In the standard textbooks Newton's third law of motion is explained with the help of above examples. But the law will be quantitatively justified if exact calculations are made.

   (a) **Swimming of a person:** Consider a person weighing 80 kg is swimming in a pool of water and moving with a velocity of 2 m/s. The swimmer pushes the water backward with hands, it is regarded as action. As the reaction the water pushes the swimmer in forward direction. Then the momentum of the person must be 160 kgm/s. Thus,

The momentum of water pushed backward (action) = $mv$ where $m$ is the mass of water pushed and $v$ is the velocity with which the water is pushed.

The momentum of the swimmer (reaction)
$= 80 \text{ kg} \times 2 \text{ m/s} = 160 \text{ kgm/s}$

Due to this action and reaction, the swimmer moves in the forward direction. According to the third law of motion action and reaction are equal and opposite.

The momentum of water pushed backward (action) = Momentum of the swimmer (reaction)

or $$mv = 160 \text{ kg m/s} \tag{10.42}$$

The backward velocity of water can be calculated from Eq. (10.42) if the mass $m$ is estimated..

(b) **Movement of a rocket:** When the fuel of the rocket burns, the compressed gases are emitted backward. It may be regarded as action. Let $m$ be the mass of compressed gases and $v$ the velocity of the gases. As the reaction the rocket also moves in the forward direction. Let us express the action and the reaction in terms of momentum.

The momentum of compressed gases (action) = $mv$

where $m$ is the mass of gases and $v$ is velocity.

Let the mass of the rocket be 1000 kg and moving in the forward direction with a velocity of 10 m/s.

The forward momentum of the rocket (reaction) = $10^4$ kgm/s

According to the third law of motion the action and reaction must be equal.

The momentum of compressed gases (action) = The forward momentum of the rocket (reaction)

rr $$mv = 10^4 \text{ kgm/s} \tag{10.43}$$

Thus, estimating the mass of the emitted gases, their velocity can be calculated.

2. **There are many equations involving elastic collisions ($e$ = 1) when the projectile (body $A$) rebounds. But such equations are not experimentally confirmed. Explain.**

**Answer:** It is very simple. The real beauty of science lies in experimentally confirming the equations rather than theoretically deriving equations. At this stage scientists have developed many sensitive techniques to measure various parameters. So it is high time to conduct the experiments and understand the laws in a more authentic and scientific way.

3. **Are the collisions discussed above elastic in nature?**
**Answer:** Yes, the collisions are elastic in nature and the coefficient of restitution is unity and based upon the equations of conservation of linear momentum and kinetic energy. Thus, all conditions for elastic collisions are satisfied.

4. **What is role of acceleration in the elastic collisions?**
Answer: The final velocities of the projectile and the target are uniform, hence the bodies posses no acceleration. Also, the acceleration does not appear in any equation at all. Thus the motion in the elastic collision is regarded as non-accelerated ($a = 0$). Hence the kinematic equations

$$v = u + at,\ v^2 - u^2 = 2aS,\ S = ut + \frac{1}{2}at^2$$

are not applicable.

## References


1. Newton, I., *Mathematical Principles of Natural Philosophy,* pp. 19–20, London, 1727, translated by Andrew Motte from the Latin.
2. Sharma, A., Acta Ciencia Indica P, Vol., XXV P, No. 3, 113 (1999).

# Index

## A



## B



## C

## M



## N



## O



## P

## R



## S



## T



## U



## V